2016: An Astronomical Year

A Reference Guide to 365 Nights of Astronomy

Richard J. Bartlett

Copyright © 2014, 2015 by Richard J. Bartlett

Published by Stars 'n Stuff Publishing

All rights reserved. Except for brief excerpts used in a review, this book or any portion thereof may not be reproduced or used in any manner whatsoever without the express written permission of the publisher.

Second Edition, August 2015

Cover image: An astronomical conjunction over Paranal. Credit: ESO/B. Tafreshi

Contents

Introduction 5

2016 11

    Solar System Summary 12

    January 21

    February 29

    March 35

    April 45

    May 53

    June 61

    July 69

    August 77

    September 85

    October 93

    November 101

    December 109

Introduction

Who This Book Is For
This book is not for the absolute beginner. If you're new to the hobby, I would recommend joining a local astronomy club or any number of astronomy groups online; there are always like-minded individuals who are willing and able to help and encourage anyone just starting out in astronomy.

Likewise, there are a multitude of books available through Amazon and other retailers to help you get started.

If, however, you have a little experience and know your conjunctions from your oppositions and your magnitude from your apparent diameter, then this book is probably just fine for you.

How to Use This Book
The book is divided into five sections, one for each year. Each section begins with a summary of moon phases, eclipses, planetary visibility, notable conjunctions and meteor showers for the year.

The book then details each individual month with notes on what's happening and what's visible on each of the key dates in that month. The text is designed to be short and to the point and aims to provide the most relevant information in the least amount of time.

Here's an example:

January 11th

01:00 UT - Conjunction of Mercury and Venus. Separation: 38.5'. (Mercury: 71% illuminated, magnitude -0.7, diameter 6.2". Venus: 95% illuminated, magnitude -3.9, diameter 10.5". Capricornus, evening sky.)

All the times are given in Universal Time (UT), the standard "time zone" used in the astronomical community. If you live in the United Kingdom, then the only conversion you'll need to add is an hour for British Summer Time.

Unfortunately, for the rest of the world, it's just not practical to convert all the times into the different time zones that are in force across the globe, but you can find an easy conversion tool at the following URL: http://www.worldtimeserver.com/convert_time_in_UTC.aspx

Next is a description of the event itself with details given for each planet or asteroid involved. You'll note that I don't specify the angular separation for conjunctions involving the Moon; this is because the Moon's separation from a star or planet will differ depending upon your latitude. Therefore, for example, the Moon will often appear to occult Aldebaran from some locations but not from others.

Bear in mind that the timing of lunar occultations are also subject to your longitude as the event may take place during daylight at your location. So even though the Moon might be visible in the evening sky, it may have moved on slightly from the time of the event itself. (It will still be worth observing but the conjunction may not appear as close.)

Fortunately, planets move slowly enough to make purely planetary conjunctions observable for several days, both before and after the date of closest approach.

At the end of the description the text will tell you if the object(s) are visible in the pre-dawn or evening sky or not visible at all.

Acknowledgements
The vast majority of the data in this book comes from a single source, the *Mobile Observatory* app for Android devices by Wolfgang Zima. I bought this excellent app a few years ago and I've used it almost daily ever since. Quite honestly, I'd recommend it to anyone and everyone with an Android device and I want to thank Wolfgang for allowing me to use the data. You can download it by searching for Mobile Observatory from the Google Play store or by visiting zima.co online.

I'd also like to thank the members of the *Telescope Addicts – Astronomy & Astrophotography Community* on Facebook. It was their generous support that helped to make *2015 An Astronomical* year such a success in the first place.

About the Author
I've had an interest in astronomy since I was six and although my interest has waxed and waned like the Moon, I've always felt compelled to stop and stare at the stars.

In the late 90's, I discovered the booming frontier of the internet, and like a settler in the Midwest, I quickly staked my claim on it. I started to build a (now-defunct) website called *StarLore*. It was designed to be an online resource for amateur astronomers who wanted to know more about the constellations - and all the stars and deep sky objects to be found within them. It was quite an undertaking.

After the website was featured in the February 2001 edition of *Sky & Telescope* magazine, I began reviewing astronomical websites and software for their rival, *Astronomy*. This was something of a dream come true; I'd been reading the magazine since I was a kid and now my name was regularly appearing in it.

Unfortunately, a financial downturn forced my monthly column to be cut after a few years but I'll always be grateful for the chance to write for the world's best-selling astronomy magazine.

I emigrated from England to the United States in 2004 and spent three years under relatively clear, dark skies in Oklahoma. I then relocated to Kentucky in 2008 and then California in 2013. I now live in the suburbs of Los Angeles; not the most ideal location for astronomy, but there are still a number of naked eye events that are easily visible on any given night.

Also by the Author…
2016 An Astronomical Year is also available as a Kindle edition in the United States, Canada and the United Kingdom. (As the Kindle format is

electronic and there are no printing costs, there are full color images to accompany the text.)

2016 The Night Sky Sights is specifically designed for absolute beginners and casual stargazers without a telescope. The guide highlights over 125 astronomical events in 2016 - all of them visible with just your eyes - and showcases events visible in both the evening and pre-dawn sky as well as those you can see throughout the night.

It is currently available in paperback and Kindle editions in the United States, Canada and the United Kingdom.

The Astronomical Almanac (2016-2020): A Comprehensive Guide to Night Sky Events provides details of thousands of astronomical events from 2016 to the end of 2020. Designed for more experience astronomers, this the guide includes almost daily data and information on the Moon and planets, as well as Pluto, Ceres, Pallas, Juno and Vesta.

To date, the 2015-2019 edition has been downloaded nearly 6,000 times, was ranked #1 in the Free Kindle Astronomy book category, #3 in the Paid Kindle Astronomy book category and within the Top 50 of *all* Free Kindle books in October 2014.

It is available in paperback and Kindle editions worldwide, including the United States, Canada, the United Kingdom and Australia.

The Amateur Astronomer's Notebook: A Journal for Recording and Sketching Astronomical Observations is the perfect way to log your observations of the Moon, stars, planets and deep sky objects. It is available as both a full-size 8.5" by 11" journal and also as a 5" by 8" pocket notebook. The larger edition has room for 150 observing sessions while the pocket edition allows you to record 100 observations.

It is available as a paperback as in the United States, Canada and the United Kingdom.

Echoes of Earth – a collection of science fiction, mythological and philosophical short stories that I wrote many, many moons ago. (i.e., in the mid 1990's.)

It is available as a Kindle edition in selected areas. (United States, Canada, the United Kingdom and Australia.)

The Author Online

Email: astronomywriter@gmail.com

Amazon US: http://tinyurl.com/rjbamazon-us

Amazon UK: http://tinyurl.com/rjbamazon-uk

The Astronomical Year: http://tinyurl.com/theastroyear

Facebook: http://tinyurl.com/rjbfacebook

Twitter (@astronomywriter): http://tinyurl.com/rjbtwitter

Clear skies!

Richard J. Bartlett

September, 2014

2016

Solar System Summary

Moon

Lunar Phases

	Jan	Feb	Mar	Apr	May	Jun	Jul	Aug	Sep	Oct	Nov	Dec
1st		◐	◐						●	●		
2nd	◐						●					
3rd												
4th						●						
5th					●							
6th				●								
7th				●							◐	◐
8th		●										
9th			◐						◐	◐		
10th	●							◐				
11th												
12th						◐	◐					
13th					◐							
14th				◐							○	○
15th		◐	◐									

Lunar Phases (cont)

	Jan	Feb	Mar	Apr	May	Jun	Jul	Aug	Sep	Oct	Nov	Dec
16th	◐								○	○		
17th												
18th								○				
19th							○					
20th						○						
21st					○						◑	◑
22nd		○		○						◑		
23rd			○						◑			
24th	○											
25th								◑				
26th							◑					
27th						◑						
28th												
29th					◑						●	●
30th				◑								
31st			◑									

13

Eclipses

Lunar Eclipses

March 23rd - 11:48 UT – Penumbral lunar eclipse. Visible from Antarctica, the Arctic, most of Asia, the Atlantic, Australia, the Indian Ocean, most of North and South America and the Pacific.

September 16th - 18:55 UT – Penumbral lunar eclipse. Visible from Africa, Antarctica, the Arctic, most of Asia, the Atlantic, Australia, Europe, the Indian and Pacific oceans and western South America.

Solar Eclipses

March 9th - 02:00 UT – Total solar eclipse. Visible from south and eastern Asia, north and eastern Australia, the Indian and Pacific oceans.

September 1st - 09:02 UT – Annular solar eclipse. Visible from most of Africa, Antarctic, south Asia, the Atlantic, eastern Australia and the Indian Ocean.

Planet Visibility

January to April

		Mer	Ven	Mar	Jup	Sat	Ura	Nep	Plu	Cer	Pal	Jun	Ves
January	Early	■							C		■		C
	Mid	C							■		C		C
	Late	■							■		■		
February	Early	E						■	■		■		
	Mid							■	■		■		
	Late							■	C		■		
March	Early	■			O			■			C		
	Mid	■					■	■			■		
	Late	C					■	■			■		
April	Early	■					■				■		
	Mid	E	■				■	C			■		
	Late	■					■				■		O

Key					
	☐	Evening Sky		C	In conjunction with the Sun
	☐	Pre-dawn sky		E	Greatest elongation (Mercury & Venus only)
	■	Not Visible		O	Opposition with the Sun. (Outer planets and asteroids only)

May to August

		Mer	Ven	Mar	Jup	Sat	Ura	Nep	Plu	Cer	Pal	Jun	Ves
May	Early	C											
	Mid												
	Late			O									
June	Early	E	C				O						
	Mid												
	Late												
July	Early	C											
	Mid								O				
	Late												
August	Early												
	Mid	E									O		
	Late												

Key		Evening Sky		C	In conjunction with the Sun
		Pre-dawn sky		E	Greatest elongation (Mercury & Venus only)
		Not Visible		O	Opposition with the Sun. (Outer planets and asteroids only)

September to December

		Mer	Ven	Mar	Jup	Sat	Ura	Nep	Plu	Cer	Pal	Jun	Ves
September	Early	■			■			O					
	Mid	C			■								
	Late	E			C								
October	Early				■								O
	Mid				■		O						
	Late	C								O		■	
November	Early	■											
	Mid	■											
	Late	■				■							
December	Early					■						C	
	Mid	E				C			■			■	
	Late								■			■	

Key		Evening Sky		C	In conjunction with the Sun
		Pre-dawn sky		E	Greatest elongation (Mercury & Venus only)
		Not Visible		O	Opposition with the Sun. (Outer planets and asteroids only)

Notable Conjunctions

Pre-dawn sky

January 9th - 03:52 UT – Venus is 6' north of Saturn. (Venus: 80% illuminated, magnitude -4.0, diameter 13.7". Saturn: magnitude 0.5, diameter 15.3". Ophiuchus, pre-dawn sky.)

August 25th - 16:57 UT – The last quarter Moon is south of Aldebaran. An occultation will be visible from some parts of the world. (Taurus, pre-dawn sky.)

Evening sky

February 16th - 07:26 UT – The just-past first quarter Moon is north of Aldebaran. An occultation will be visible from some parts of the world. (Taurus, evening sky.)

July 30th - 14:36 UT – Mercury is 20' north of Regulus. (Mercury: 74% illuminated, magnitude -0.2, diameter 5.8". Leo, evening sky.)

August 27th - 22:46 UT – Venus is 4' north of Jupiter. (Venus: 92% illuminated, magnitude -3.9, diameter 10.8". Jupiter: magnitude -1.7, diameter 30.9". Virgo, evening sky.)

September 21st – 20:43 - 20:43 UT – The waning gibbous Moon is south of Aldebaran. An occultation may be visible from some parts of the world. (Taurus, pre-dawn sky.)

Meteor Showers

January 4th - The Quadrantid meteor shower peaks. Maximum zenith hourly rate: 120. (Moon: waning crescent. Boötes.)

April 22nd - The Lyrid meteor shower peaks. Maximum zenith hourly rate: 18. (Moon: full. Lyra.)

May 5th - The Eta Aquariid meteor shower peaks. Maximum zenith hourly rate: 65. (Moon: almost new. Aquarius.)

June 7th - The Arietid meteor shower peaks. Maximum zenith hourly rate: 54. (Moon: waxing crescent. Aries.)

July 29th - The Southern Delta Aquariid meteor shower peaks. Maximum zenith hourly rate: 16. (Moon: waning crescent. Aquarius.)

July 29th - The Beta Cassiopeid meteor shower peaks. Maximum zenith hourly rate: 10. (Moon: waning crescent. Cassiopeia.)

August 12th - The Perseid meteor shower peaks. Maximum zenith hourly rate: 100. (Moon: waxing gibbous. Perseus.)

October 8th - The Draconid meteor shower peaks. Maximum zenith hourly rate: Variable. (Moon: almost first quarter. Draco.)

October 21st - The Orionid meteor shower peaks. Maximum zenith hourly rate: 25. (Moon: almost last quarter. Orion.)

November 17th - The Leonid meteor shower peaks. Maximum zenith hourly rate: 15. (Moon: waning gibbous. Leo.)

December 13th - The Geminid meteor shower peaks. Maximum zenith hourly rate: 120. (Moon: almost full. Gemini.)

December 23rd - The Ursid meteor shower peaks. Maximum zenith hourly rate: 10. (Moon: waning crescent. Ursa Minor.)

January

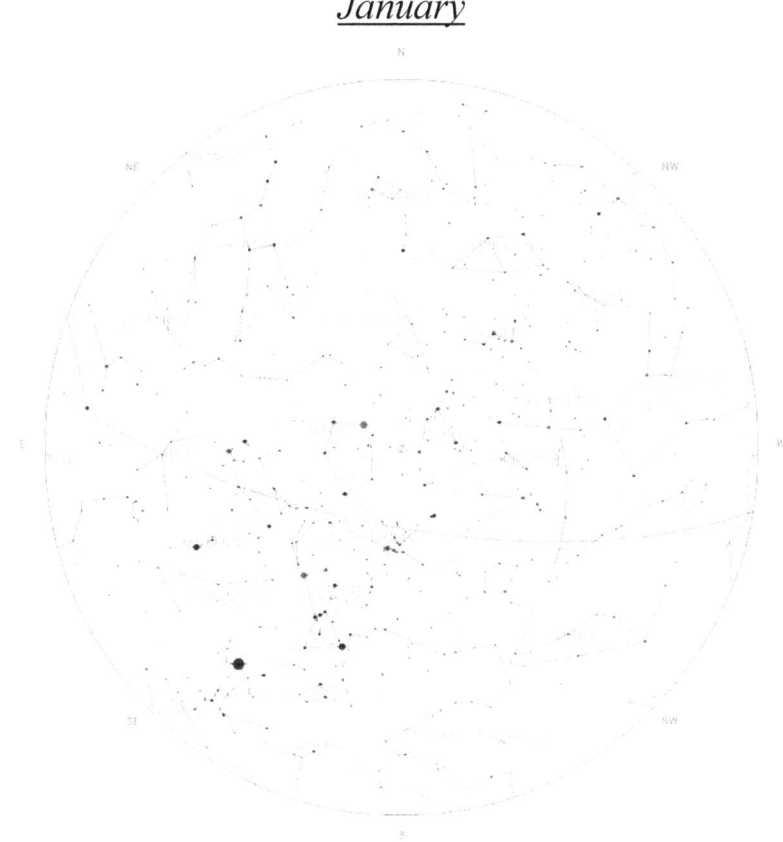

This image depicts the sky at 10:00 p.m. on the 1st but the positions of the stars will be the same at 9:00 p.m. on the 15th and 8:00 p.m. on the 31st.

Phases of the Moon

Last Quarter	New Moon	First Quarter	Full Moon
January 2nd	January 10th	January 16th	January 24th

January 1st

Mercury leaves Sagittarius and enters Capricornus. (45% illuminated, magnitude -0.2, diameter 7.5", evening sky.)

Venus leaves Libra and enters Scorpius. (77% illuminated, magnitude -4.1, diameter 14.3", pre-dawn sky.)

January 2nd

05:30 UT – Last Quarter Moon (Virgo, pre-dawn sky.)

20:24 UT - The Earth is at perihelion. Distance to Sun: 0.983 AU

January 3rd

01:46 UT – The just-past last quarter Moon is north of Spica. (Virgo, pre-dawn sky.)

02:38 UT - Summer begins in the northern hemisphere of Mars. (91% illuminated, magnitude 1.2, diameter 5.6". Virgo, pre-dawn sky.)

11:00 UT - Mercury fades to magnitude 0.0. (38% illuminated, diameter 7.8". Capricornus, evening sky.)

20:24 UT – Dwarf planet Ceres is at aphelion. Distance to Sun: 3.683 AU. (Ceres: magnitude 8.9. Capricornus, evening sky.)

18:52 UT – The waning crescent Moon is north of Mars. (Mars: 91% illuminated, magnitude 1.2, diameter 5.6". Virgo, pre-dawn sky.)

January 4th

The Quadrantid meteor shower peaks. Maximum zenith hourly rate: 120. (Moon: waning crescent. Boötes.)

January 5th

04:38 UT - Mercury is stationary prior to beginning retrograde motion. (29% illuminated, magnitude 0.4, diameter 8.2". Capricornus, evening sky.)

18:10 UT – Pluto is in conjunction with the Sun. (Pluto: magnitude 14.3. Sagittarius, not visible.)

Mercury fades to magnitude 0.5. (30% illuminated, diameter 8.2". Capricornus, evening sky.)

Venus leaves Scorpius and enters Ophiuchus. (78% illuminated, magnitude -4.1, diameter 14.0", pre-dawn sky.)

January 6th

12:09 UT – Venus is 6.5° north of Antares. (Venus: 79% illuminated, magnitude -4.1, diameter 13.9". Ophiuchus, pre-dawn sky.)

22:23 UT – The waning crescent Moon is north of Antares. (Ophiuchus, pre-dawn sky.)

23:07 UT – The waning crescent Moon is north of Venus. (Venus: 79% illuminated, magnitude -4.0, diameter 13.9". Ophiuchus, pre-dawn sky.)

Venus fades to magnitude -4.0. (79% illuminated, diameter 13.9". Ophiuchus, pre-dawn sky.)

Good opportunity to see Earthshine on the waning crescent Moon. (Pre-dawn sky.)

January 7th

02:52 UT – The waning crescent Moon is north of Saturn. (Saturn: magnitude 0.5, diameter 15.3", Ophiuchus, pre-dawn sky.)

Mercury fades to magnitude 1.0. (21% illuminated, diameter 8.7".
Capricornus, not visible.)

January 8th

01:33 UT – Pluto is at apogee. Distance to Earth: 34.000 AU. (Pluto: magnitude 14.3. Sagittarius, not visible.)

17:36 UT - Mercury reaches perihelion. Distance from the Sun: 0.308 AU. (13% illuminated, magnitude 1.6, diameter 9.1". Capricornus, not visible.)

19:12 UT – Jupiter is stationary prior to beginning retrograde motion. (Magnitude -2.2, diameter 39.9", Leo, pre-dawn sky.)

Mercury leaves Capricornus and returns to Sagittarius. (16% illuminated, magnitude 1.3, diameter 8.9", evening sky.)

January 9th

03:52 UT – Venus is 6' north of Saturn. (Venus: 80% illuminated, magnitude -4.0, diameter 13.7". Saturn: magnitude 0.5, diameter 15.3". Ophiuchus, pre-dawn sky.)

20:18 UT – The almost new Moon is north of Pluto. (Pluto: magnitude 14.3. Sagittarius, not visible.)

Jupiter increases its apparent diameter to 40.0". (Magnitude -2.2, Leo, pre-dawn sky.)

January 10th

01:30 UT – New Moon (Sagittarius, not visible.)

19:52 UT – The new Moon is north of Mercury. (6% illuminated, magnitude 2.7, diameter 9.5". Sagittarius, not visible.)

January 12th

21:18 UT – Asteroid 2 Pallas is in conjunction with the Sun. (Pallas: Magnitude 10.2. Aquila, not visible.)

Good opportunity to see Earthshine on the waxing crescent Moon. (Evening sky.)

January 13th

14:23 UT – The waxing crescent Moon is north of Neptune. (Neptune: magnitude 7.9, diameter 2.2". Aquarius, evening sky.)

Mercury increases its apparent diameter to 10.0". (2% illuminated, magnitude 4.1. Sagittarius, not visible.)

January 14th

13:59 UT - Mercury is at inferior conjunction with the Sun. Distance to Earth: 0.668 AU. (1% illuminated, magnitude 4.7, diameter 10.0". Sagittarius, not visible.)

January 15th

15:24 UT - Mercury is at perigee. Distance to Earth: 0.667 AU. (1% illuminated, magnitude 4.4, diameter 10.0". Sagittarius, not visible.)

Dwarf planet Ceres leaves Capricornus and enters Aquarius. (Magnitude 8.9, evening sky.)

January 16th

06:00 UT – The almost first quarter Moon is south of Uranus. (Uranus: magnitude 5.8, diameter 3.5". Pisces, evening sky.)

23:26 UT - First Quarter Moon. This is the largest first quarter Moon of the year with an apparent diameter of 32.234'. (Pisces, evening sky.)

January 17th

Mars leaves Virgo and enters Libra. (91% illuminated, magnitude 1.1, diameter 6.1", pre-dawn sky.)

Mars brightens to magnitude 1.0. (93% illuminated, diameter 6.1". Libra, pre-dawn sky.)

January 19th

06:21 UT – The waxing gibbous Moon is south of M45, the Pleiades open star cluster. (Taurus, evening sky.)

January 20th

03:54 UT – The waxing gibbous Moon is north of Aldebaran (Taurus, evening sky.)

The Sun leaves Sagittarius and enters Capricornus.

Venus leaves Ophiuchus and enters Sagittarius. (82% illuminated, magnitude -4.0, diameter 13.0", pre-dawn sky.)

Neptune fades to magnitude 8.0. (Diameter 2.2". Aquarius, evening sky.)

January 22nd

10:25 UT – Mercury is 1.7° south north of Pluto. (19% illuminated, magnitude 1.3, diameter 9.3". Pluto: magnitude 14.2. Sagittarius, not visible.)

January 23rd

10:54 UT – The almost full Moon is south of Pollux. (Gemini, visible all night.)

Mercury brightens to magnitude 1.0. (21% illuminated, diameter 9.2". Sagittarius, pre-dawn sky.)

January 24th

01:45 UT - Full Moon (Cancer, visible all night.)

11:32 UT – The full Moon is south of M44, the Praesepe open star cluster. (Cancer, visible all night.)

23:40 UT – Asteroid 2 Pallas is at apogee. Distance to Earth: 4.284 AU. (Magnitude 10.2. Aquila, not visible.)

January 25th

18:39 UT – Mercury is stationary prior to resuming prograde motion. (30% illuminated, magnitude 0.6, diameter 8.6". Sagittarius, pre-dawn sky.)

January 26th

06:59 UT – The waning gibbous Moon is south of Regulus. (Leo, pre-dawn.)

Mercury brightens to magnitude 0.5. (31% illuminated, diameter 8.6". Sagittarius, pre-dawn sky.)

January 27th

23:52 UT – The waning gibbous Moon is south of Jupiter. (Jupiter: magnitude -2.3, diameter 42.0", Leo, pre-dawn sky.)

Asteroid 3 Juno leaves Virgo and enters Libra (Magnitude 10.7, pre-dawn sky.)

January 28th

Uranus fades to magnitude 5.9. (Apparent diameter 3.5". Pisces, evening sky.)

January 30th

13:09 UT – The waning gibbous Moon is north of Spica. (Virgo, pre-dawn sky.)

February

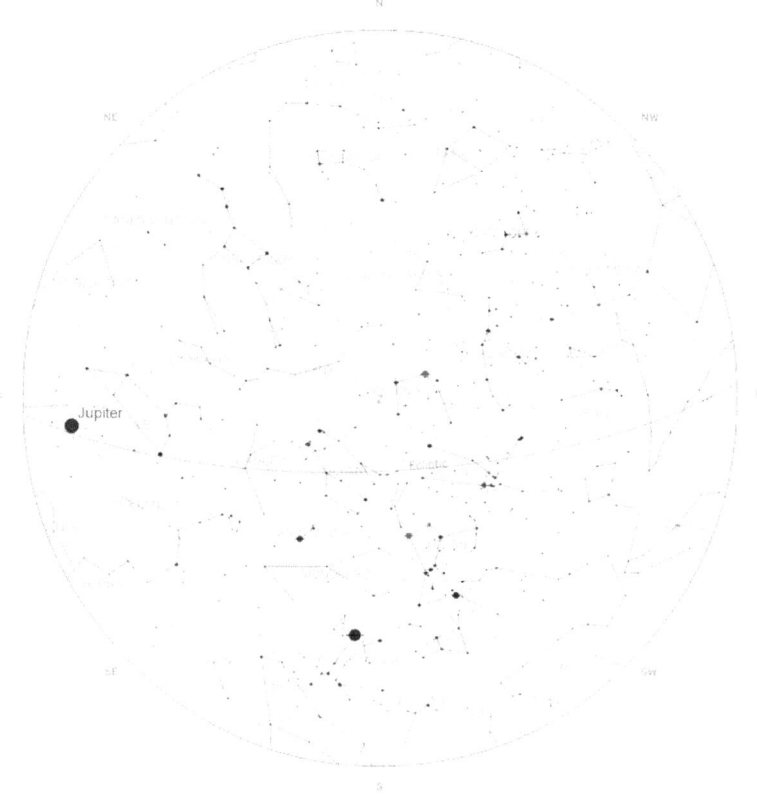

This image depicts the sky at 10:00 p.m. on the 1st but the positions of the stars will be the same at 9:00 p.m. on the 15th and 8:00 p.m. on the 29th.

Phases of the Moon

Last Quarter	New Moon	First Quarter	Full Moon
February 1st	February 8th	February 15th	February 22nd

February 1st

03:27 UT - Last Quarter Moon (Libra, pre-dawn sky.)

10:38 UT – The last quarter Moon is north of Mars. (Mars: 90% illuminated, magnitude 0.8, diameter 6.8". Libra, pre-dawn sky.)

12:24 UT - Mercury is 50% illuminated. (Magnitude 0.1, diameter 7.5". Sagittarius, pre-dawn sky.)

February 3rd

08:09 UT – The waning crescent Moon is north of Antares. (Ophiuchus, pre-dawn sky.)

19:09 UT – The waning crescent Moon is north of Saturn. (Saturn: magnitude 0.6, diameter 15.8", Ophiuchus, pre-dawn sky.)

February 4th

21:00 UT - Mercury brightens to magnitude 0.0. (51% illuminated, diameter 7.4". Sagittarius, pre-dawn sky.)

Good opportunity to see Earthshine on the waning crescent Moon. (Pre-dawn sky.)

February 5th

18:16 UT – Venus is 1.1° south of Pluto. (Venus: 86% illuminated, magnitude -4.0, diameter 12.1". Pluto: magnitude 14.2. Sagittarius, pre-dawn sky.)

February 6th

04:37 UT – The waning crescent Moon is north of Pluto. (Pluto: magnitude 14.2. Sagittarius, pre-dawn sky.)

05:51 UT – The waning crescent Moon is north of Venus. (Venus: 86% illuminated, magnitude -4.0, diameter 12.1". Sagittarius, pre-dawn sky.)

18:25 UT – The waning crescent Moon is north of Mercury. (Mercury: 61% illuminated, magnitude -0.0, diameter 6.8". Sagittarius, pre-dawn sky.)

February 7th

01:14 UT - Mercury is at Greatest Western Elongation. (62% illuminated, magnitude -0.0, diameter 6.7". Sagittarius, pre-dawn sky.)

February 8th

14:39 UT - New Moon (Aquarius, not visible.)

February 9th

Asteroid 3 Juno brightens to magnitude 10.5. (Libra, pre-dawn sky.)

February 10th

00:40 UT – The waxing crescent Moon is north of Neptune. (Neptune: magnitude 8.0, diameter 2.2". Aquarius, not visible.)

Saturn increases its apparent diameter to 16.0". (Magnitude 0.5, Ophiuchus, pre-dawn sky.)

February 11th

Good opportunity to see Earthshine on the waxing crescent Moon. (Evening sky.)

February 12th

12:29 UT – The waxing crescent Moon is south of Uranus. (Uranus: magnitude 5.9, diameter 3.4". Pisces, evening sky.)

Asteroid 4 Vesta leaves Cetus and enters Pisces. (Magnitude 7.9, evening sky.)

February 13th

Mercury leaves Sagittarius and returns to Capricornus. (71% illuminated, magnitude -0.1, diameter 6.2". Morning sky.)

February 15th

07:46 UT - First Quarter Moon (Taurus, evening sky.)

09:36 UT – The first quarter Moon is south of M45, the Pleiades open star cluster. (Taurus, evening sky.)

February 16th

07:26 UT – The just-past first quarter Moon is north of Aldebaran. An occultation will be visible from some parts of the world. (Taurus, evening sky.)

The Sun leaves Capricornus and enters Aquarius.

Venus leaves Sagittarius and enters Capricornus. (88% illuminated, magnitude -3.9, diameter 11.7", pre-dawn sky.)

Mars brightens to magnitude 0.5. (90% illuminated, diameter 7.7". Libra, pre-dawn sky.)

Asteroid 4 Vesta fades to magnitude 8.0. (Pisces, evening sky.)

February 18th

Asteroid 2 Pallas leaves Aquila and enters Delphinus. (Magnitude 10.2, pre-dawn sky.)

February 19th

15:37 UT – The waxing gibbous Moon is south of Pollux. (Gemini, evening sky.)

Jupiter brightens to magnitude -2.5 (Diameter 43.8", Leo, evening sky.)

February 20th

16:33 UT – The waxing gibbous Moon is south of M44, the Praesepe open star cluster. (Cancer, evening sky.)

February 21st

17:14 UT - Mercury is at aphelion. Distance to Sun: 0.467 AU. (80% illuminated, magnitude -0.1, diameter 5.6". Capricornus, pre-dawn sky.)

February 22nd

12:33 UT – The almost full Moon is south of Regulus (Leo, visible all night.)

18:20 UT - Full Moon (Leo, visible all night.)

February 24th

05:48 UT – The waning gibbous Moon is south of Jupiter. (Jupiter: magnitude -2.5, diameter 44.1", Leo, pre-dawn sky.)

February 26th

17:49 UT – The waning gibbous Moon is north of Spica. (Virgo, evening sky.)

February 28th

23:51 UT – Neptune is in conjunction with the Sun. Distance to Earth: 30.949 AU. (Magnitude 8.0, diameter 2.2". Aquarius, not visible.)

February 29th

10:25 UT – Neptune is at apogee. Distance to Earth: 30.949 AU. (Magnitude 8.0, diameter 2.2". Aquarius, not visible.)

17:54 UT – The waning gibbous Moon is north of Mars. (Mars: 90% illuminated, magnitude 0.2, diameter 8.6". Libra, pre-dawn sky.)

March

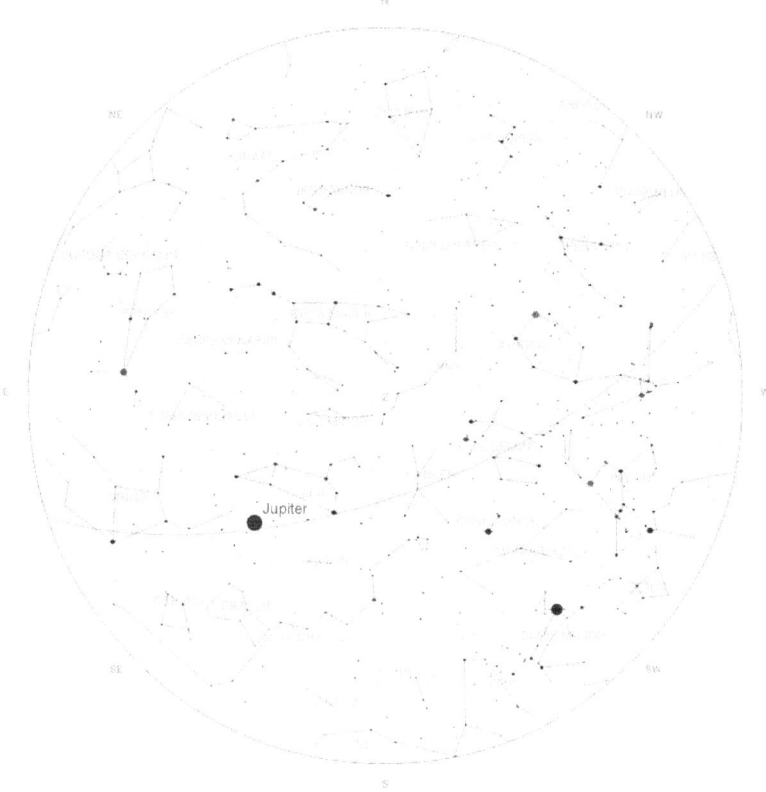

This image depicts the sky at 11:00 p.m. on the 1st but the positions of the stars will be the same at 11:00 p.m. (Daylight Savings) on the 15th and 10:00 p.m. on the 31st.

Phases of the Moon

Last Quarter	New Moon	First Quarter	Full Moon	Last Quarter
March 1st	March 9th	March 15th	March 23rd	March 31st

March 1st

16:45 UT – The almost last-quarter Moon is north of Antares. (Ophiuchus, pre-dawn sky.)

23:10 UT - Last Quarter Moon (Ophiuchus, pre-dawn sky.)

March 2nd

06:47 UT – The waning crescent Moon is north of Saturn. (Saturn: magnitude 0.5, diameter 16.5", Ophiuchus, pre-dawn sky.)

March 3rd

23:29 UT – Dwarf planet Ceres reaches its minimum brightness. (Magnitude 8.9. Aquarius, not visible.)

Mercury leaves Capricornus and enters Aquarius. (89% illuminated, magnitude -0.4, diameter 5.2", pre-dawn sky.)

March 4th

13:32 UT – Dwarf planet Ceres is at apogee. Distance to Earth: 3.951 AU. (Magnitude 8.9. Aquarius, not visible.)

18:27 UT – The waning crescent Moon is north of Pluto. (Pluto: magnitude 14.2. Sagittarius, pre-dawn sky.)

Mercury brightens to magnitude -0.5. (89% illuminated, diameter 5.1". Aquarius, not visible.)

March 5th

Good opportunity to see Earthshine on the waning crescent Moon. (Pre-dawn sky.)

March 6th

01:01 UT – Asteroid 3 Juno is stationary prior to beginning retrograde motion. (Magnitude 10.3. Libra, pre-dawn sky.)

Mercury decreases its apparent diameter to 5.0". (91% illuminated, magnitude -0.5. Aquarius, not visible.)

March 7th

11:06 UT – The waning crescent Moon is north of Venus. (Venus: 92% illuminated, magnitude -3.9, diameter 11.0". Capricornus, pre-dawn sky.)

March 8th

03:29 UT – The almost new Moon is north of Mercury. (Mercury: 92% illuminated, magnitude -0.6, diameter 5.0". Aquarius, not visible.)

09:55 UT – Jupiter reaches its maximum brightness. (Magnitude -2.5, diameter 44.4", Leo, visible all night.)

13:01 UT – The almost new Moon is north of Neptune. (Neptune: magnitude 8.0, diameter 2.2". Aquarius, not visible.)

18:18 UT – Jupiter is at perigee. Distance to Earth: 4.435 AU. (Magnitude -2.5, diameter 44.4", Leo, visible all night.)

23:20 UT – First location to see a partial solar eclipse begin.

March 9th

00:02 UT – Jupiter is at opposition. (Jupiter: magnitude -2.5, diameter 44.4", Leo, visible all night.)

00:16 UT – First location to see the total solar eclipse begin.

01:54 UT - New Moon (Aquarius, not visible.)

02:00 UT – Total solar eclipse maximum. Visible from south and eastern Asia, north and eastern Australia, the Indian and Pacific oceans.

03:03 UT – Dwarf planet Ceres is in conjunction with the Sun. (Ceres: magnitude 8.9. Aquarius, evening sky.)

03:39 UT – Last location to see the total solar eclipse end.

04:35 UT – Last location to see a partial solar eclipse end.

March 10th

22:02 UT – Mercury is 1.5° south of Neptune. (Mercury: 94% illuminated, magnitude -0.7, diameter 5.0". Neptune: magnitude 8.0, diameter 2.2". Aquarius, not visible.)

Venus leaves Capricornus and enters Aquarius. (92% illuminated, magnitude -3.9, diameter 10.9", pre-dawn sky.)

Mars brightens to magnitude 0.0. (91% illuminated, diameter 9.5". Libra, pre-dawn sky.)

March 11th

00:59 UT – The waxing crescent Moon is south of Uranus. (Uranus: magnitude 5.8, diameter 3.5". Pisces, evening sky.)

23:59 UT – The waxing crescent Moon is north of asteroid 4 Vesta. (Vesta: magnitude 8.1. Pisces, evening sky.)

The Sun leaves Aquarius and enters Pisces.

Mars leaves Libra and enters Scorpius. (91% illuminated, magnitude 0.0, diameter 9.6". pre-dawn sky.)

Good opportunity to see Earthshine on the waxing crescent Moon. (Evening sky.)

March 12th

13:00 UT – Mars brightens to magnitude 0.0. (91% illuminated, diameter 9.7". Scorpius, pre-dawn sky.)

March 13th

18:24 UT – The waxing crescent Moon is south of M45, the Pleiades open star cluster. (Taurus, evening sky.)

Mercury brightens to magnitude -1.0. (95% illuminated, diameter 4.9". Aquarius, not visible.)

Asteroid 4 Vesta leaves Pisces and returns to Cetus. (Magnitude 7.9, evening sky.)

March 14th

12:06 UT – The waxing crescent Moon is north of Aldebaran. (Taurus, evening sky.)

March 15th

15:00 UT - Mars increases its apparent diameter to 10.0". (91% illuminated, magnitude -0.1. Scorpius, pre-dawn sky.)

17:03 UT - First Quarter Moon. This is the northernmost first quarter moon of the year. (Taurus, evening sky.)

March 16th

Asteroid 2 Pallas leaves Delphinus and enters Equuleus. (Magnitude 10.2, pre-dawn sky.)

March 18th

00:36 UT – The waxing gibbous Moon is south of Pollux. (Gemini, evening sky.)

01:14 UT – Mercury is at apogee. (98% illuminated, magnitude -1.3, diameter 4.9". Aquarius, not visible.)

Saturn increases its apparent diameter to 17.0". (Magnitude 0.4, Ophiuchus, pre-dawn sky.)

March 19th

01:52 UT – The waxing gibbous Moon is south of M44, the Praesepe open star cluster. (Cancer, evening sky.)

Mercury leaves Aquarius and enters Pisces. (99% illuminated, magnitude -1.4, diameter 4.9", not visible.)

Mercury brightens to magnitude -1.5. (99% illuminated, diameter 4.9". Pisces, not visible.)

March 20th

04:30 UT – Spring Equinox. Spring begins in the northern hemisphere, autumn begins in the southern hemisphere.

17:17 UT – The waxing gibbous Moon is south of Regulus. (Leo, evening sky.)

13:42 UT – Venus is 30' south of Neptune. (Venus: 94% illuminated, magnitude -3.9, diameter 10.6". Neptune: magnitude 8.0, diameter 2.2". Aquarius, Neptune not visible.)

13:47 UT – Venus is at aphelion. Distance to Sun: 0.728 AU. (Venus: 94% illuminated, magnitude -3.9, diameter 10.6". Aquarius, pre-dawn sky.)

March 22nd

05:52 UT – The waxing gibbous Moon is south of Jupiter. (Jupiter: magnitude -2.5, diameter 44.1", Leo, evening sky.)

March 23rd

09:42 UT – Penumbral lunar eclipse begins.

11:48 UT – Penumbral lunar eclipse maximum. Visible from Antarctica, the Arctic, most of Asia, the Atlantic, Australia, the Indian Ocean, most of North and South America and the Pacific.

12:01 UT - Full Moon. (Virgo, visible all night.)

13:53 UT – Penumbral lunar eclipse ends.

19:57 UT - Mercury is at superior conjunction with the Sun. Distance to Earth: 1.347 AU. (100% illuminated, magnitude -1.9, diameter 5.0". Pisces, not visible.)

Mercury increases its apparent diameter to 5.0". (100% illuminated, magnitude -1.8. Pisces, not visible.)

March 24th

12:43 UT – Mercury reaches its maximum brightness. (100% illuminated, magnitude -1.9, diameter 5.0". Pisces, not visible.)

March 25th

02:26 UT – The waning gibbous Moon is north of Spica. (Virgo, pre-dawn sky.)

12:00 UT – Saturn is stationary prior to beginning retrograde motion. (Saturn: magnitude 0.4, diameter 17.1", Ophiuchus, pre-dawn sky.)

March 27th

Asteroid 3 Juno brightens to magnitude 10.0. (Libra, evening sky.)

March 28th

18:51 UT – The waning gibbous Moon is north of Mars. (Mars: 92% illuminated, magnitude -0.4, diameter 11.4". Scorpius, pre-dawn sky.)

22:26 UT – The waning gibbous Moon is north of Antares. (Ophiuchus, pre-dawn sky.)

March 29th

07:00 UT – Asteroid 2 Juno is at aphelion. Distance to Sun: 3.250 AU. (Magnitude 10.0. Libra, pre-dawn sky.)

16:13 UT – The waning gibbous Moon is north of Saturn. (Saturn: magnitude 0.4, diameter 17.3", Ophiuchus, pre-dawn sky.)

Mars brightens to magnitude -0.5. (92% illuminated, diameter 11.5". Scorpius, pre-dawn sky.)

March 30th

Asteroid 4 Vesta leaves Cetus and enters Aries. (Magnitude 8.1, evening sky.)

March 31st

15:17 UT - Last Quarter Moon. This is the southernmost last quarter moon of the year. (Sagittarius, pre-dawn sky.)

23:38 UT – Mercury is 33.5' north of Uranus. (Mercury: 94% illuminated, magnitude -1.5, diameter 5.3". Uranus: magnitude 5.9, diameter 3.3". Pisces, not visible.)

Mercury fades to magnitude -1.5. (96% illuminated, diameter 5.2". Pisces, not visible.)

April

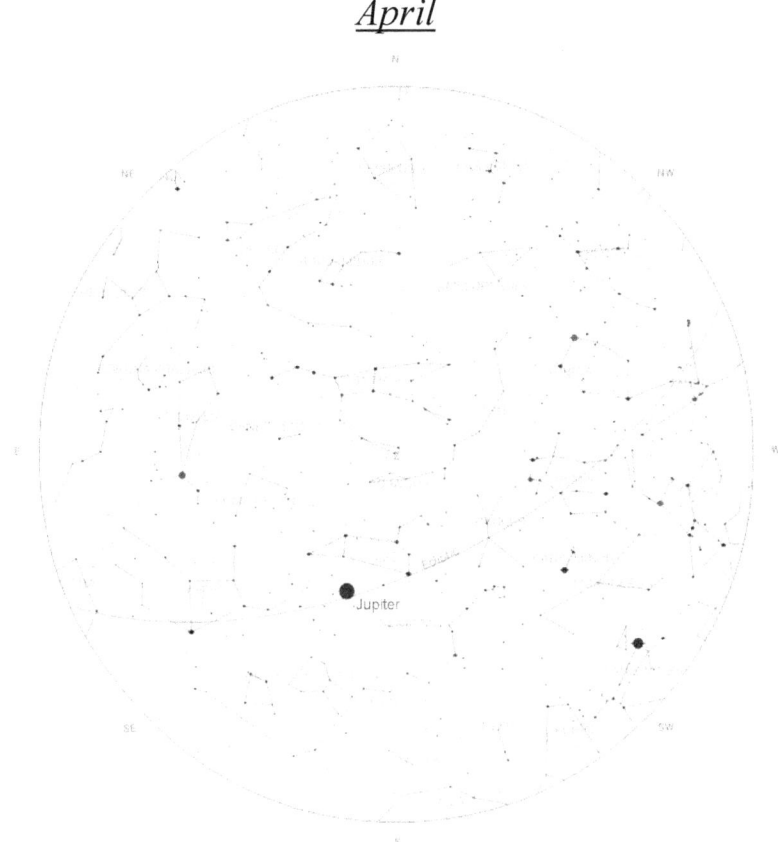

This image depicts the sky at 11:00 p.m. on the 1st but the positions of the stars will be the same at 10:00 p.m. on the 15th and 9:00 p.m. on the 30th. (Daylight Savings Time.)

Phases of the Moon

New Moon	First Quarter	Full Moon	Last Quarter
April 7th	April 14th	April 22nd	April 30th

April 1ˢᵗ

02:14 UT – The waning crescent Moon is north of Pluto. (Pluto: magnitude 14.2. Sagittarius, pre-dawn sky.)

Venus leaves Aquarius and enters Pisces. (96% illuminated, magnitude -3.9, diameter 10.3", pre-dawn sky.)

Mars leaves Scorpius and enters Ophiuchus. (93% illuminated, magnitude -0.5, diameter 11.8", pre-dawn sky.)

April 3ʳᵈ

Dwarf planet Ceres leaves Aquarius and enters Cetus. (Magnitude 8.9, not visible.)

Good opportunity to see Earthshine on the waning crescent Moon. (Pre-dawn sky.)

April 5ᵗʰ

01:00 UT – The waning crescent Moon is north of Neptune. (Neptune: magnitude 8.0, diameter 2.2". Aquarius, pre-dawn sky.)

16:52 UT - Mercury is at perihelion. Distance to Sun: 0.308 AU. (84% illuminated, magnitude -1.2, diameter 5.7". Aries, not visible.)

Mercury leaves Pisces and enters Aries. (86% illuminated, magnitude -1.3, diameter 5.6", not visible.)

April 6ᵗʰ

07:53 UT – The almost new Moon is north of Venus. (Venus: 96% illuminated, magnitude -3.9, diameter 10.2". Pisces, pre-dawn sky.)

April 7th

11:23 UT - New Moon. This is the nearest new moon of the year. Distance to Earth: 357, 233 km, 221,974 miles. (Pisces, not visible.)

15:51 UT – The new Moon is south of Uranus. Uranus: magnitude 5.9, diameter 3.3". Pisces, not visible.)

Mercury fades to magnitude -1.0. (80% illuminated, diameter 5.8". Aries, evening sky.)

April 8th

09:59 UT – The just-past new Moon is south of Mercury. (Mercury: 75% illuminated, magnitude -1.0, diameter 6.0". Aries, evening sky.)

April 9th

03:51 UT – The waxing crescent Moon is south of asteroid 4 Vesta. (Vesta: magnitude 8.1. Aries, evening sky.)

Neptune brightens to magnitude 7.9. (Diameter 2.2". Aquarius, pre-dawn sky.)

April 10th

03:34 UT – The waxing crescent Moon is south of M45, the Pleiades open star cluster. (Taurus, evening sky.)

04:42 UT – Uranus is in conjunction with the Sun. Distance to Earth: 20.968 AU. Uranus: magnitude 5.9, diameter 3.3". Pisces, not visible.)

14:54 UT - Uranus is at apogee. Distance to Earth: 20.968 AU. Uranus: magnitude 5.9, diameter 3.3". Pisces, not visible.)

Good opportunity to see Earthshine on the waxing crescent Moon. (Evening sky.)

April 11th

00:29 UT – The waxing crescent Moon is north of Aldebaran (Taurus, evening sky.)

April 12th

Mercury fades to magnitude -0.5. (62% illuminated, diameter 6.5". Aries, evening sky.)

April 14th

03:59 UT - First Quarter Moon. (Gemini, evening sky.)

06:24 UT – The first quarter Moon is south of Pollux. (Gemini, evening sky.)

Asteroid 2 Pallas leaves Equuleus and enters Pegasus. (Magnitude 10.1, pre-dawn sky.)

April 15th

07:16 UT – The just-past first quarter Moon is south of M44, the Praesepe open star cluster. (Cancer, evening sky.)

22:00 UT – Mercury is 50% illuminated. (Magnitude -0.2, diameter 7.2". Aries, evening sky.)

Asteroid 3 Juno leaves Libra and returns to Virgo (Magnitude 9.9, evening sky.)

April 16th

06:00 UT - Mars brightens to magnitude -1.0. (95% illuminated, diameter 13.9". Ophiuchus, pre-dawn sky.)

Venus decreases its apparent diameter to 10.0". (97% illuminated, magnitude -3.9. Pisces, pre-dawn sky.)

Asteroid 2 Pallas brightens to magnitude 10.0. (Pegasus, pre-dawn sky.)

April 17th

03:09 UT – Mars is stationary prior to beginning retrograde motion. (95% illuminated, magnitude -1.0, diameter 14.1". Ophiuchus, pre-dawn sky.)

03:38 UT – The waxing gibbous Moon is south of Regulus. (Leo, evening sky.)

06:00 UT - Mercury fades to magnitude 0.0. (43% illuminated, diameter 7.4". Aries, evening sky.)

April 18th

03:31 UT – Pluto is stationary prior to beginning retrograde motion. (Pluto: magnitude 14.2. Sagittarius, pre-dawn sky.)

07:09 UT – The waxing gibbous Moon is south of Jupiter. (Jupiter: magnitude -2.3, diameter 42.1", Leo, evening sky.)

13:53 UT – Mercury is at Greatest Eastern Elongation. (39% illuminated, magnitude 0.2, diameter 7.7". Aries, evening sky.)

The Sun leaves Pisces and enters Aries.

April 20th

Mercury fades to magnitude 0.5. (34% illuminated, diameter 8.0". Aries, evening sky.)

April 21st

10:05 UT – The almost full Moon is north of Spica. (Virgo, visible all night.)

April 22nd

05:23 UT - Full Moon (Virgo, visible all night.)

15:14 UT – Venus is 48' south of Uranus. (Venus: 98% illuminated, magnitude -3.9, diameter 9.9". Uranus: magnitude 5.9, diameter 3.3". Pisces, not visible.)

The Lyrid meteor shower peaks. Maximum zenith hourly rate: 18. (Moon: full. Lyra.)

April 23rd

Mercury fades to magnitude 1.0. (25% illuminated, diameter 8.7". Aries, evening sky.)

Mars increases its apparent diameter to 15.0". (96% illuminated, magnitude -1.2. Ophiuchus, evening sky.)

April 24th

11:36 UT – Asteroid 3 Juno is at perigee. (Magnitude 9.8. Virgo, visible all night.)

April 25th

06:41 UT – The waning gibbous Moon is north of Mars. (Mars: 97% illuminated, magnitude -1.3, diameter 15.3". Ophiuchus, pre-dawn sky.)

07:59 UT – The waning gibbous Moon is north of Antares. (Ophiuchus, pre-dawn sky.)

11:32 UT – Asteroid 3 Juno reaches its maximum brightness. (Magnitude 9.8. Virgo, visible all night.)

18:47 UT – The waning gibbous Moon is north of Saturn. (Saturn: magnitude 0.2, diameter 17.9", Ophiuchus, pre-dawn sky.)

April 26th

Saturn increases its apparent diameter to 18.0". (Magnitude 0.2, Ophiuchus, pre-dawn sky.)

April 27th

Mercury increases its apparent diameter to 10.0". (15% illuminated, magnitude 1.8. Aries, not visible.)

April 29th

03:27 UT – Mercury is stationary prior to beginning retrograde motion. (11% illuminated, magnitude 2.3, diameter 10.2". Aries, not visible.)

Asteroid 4 Vesta leaves Aries and enters Taurus. (Magnitude 8.1, evening sky.)

April 30th

03:28 UT - Last Quarter Moon (Capricornus, pre-dawn sky.)

22:18 UT – Asteroid 3 Juno is at opposition. (Magnitude 9.8. Virgo, visible all night.)

Venus leaves Pisces and enters Aries. (99% illuminated, magnitude -3.9, diameter 9.8", not visible.)

Mars brightens to magnitude -1.5. (98% illuminated, diameter 15.9". Ophiuchus, pre-dawn sky.)

May

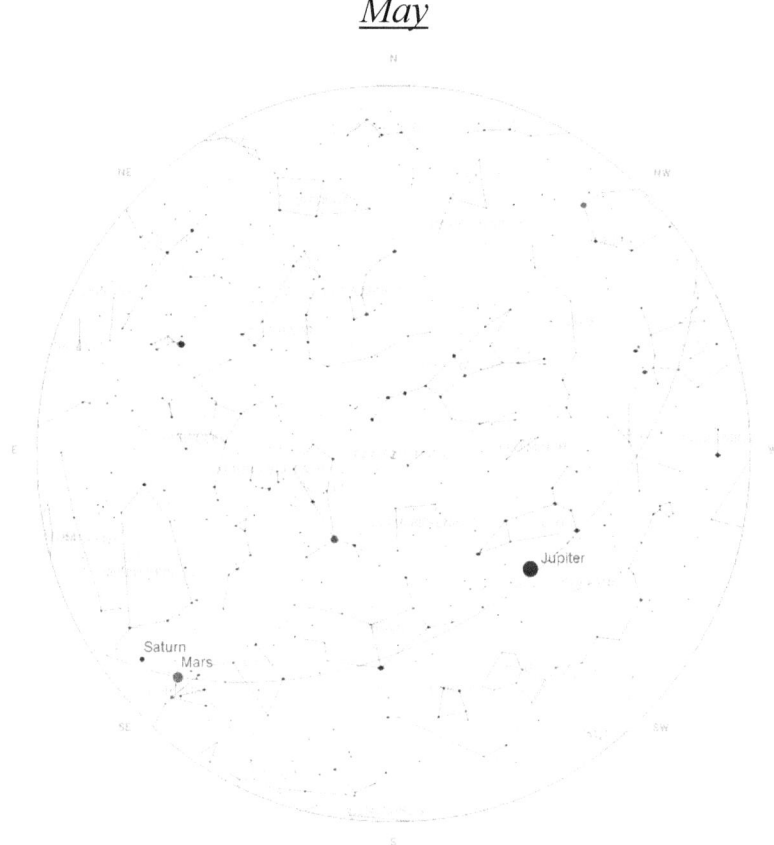

This image depicts the sky at midnight on the 1st but the positions of the stars will be the same at 11:00 p.m. on the 15th and 10:00 p.m. on the 31st. (Daylight Savings Time)

Phases of the Moon

New Moon	First Quarter	Full Moon	Last Quarter
May 6th	May 13th	May 21st	May 29th

May 1st

Mars leaves Ophiuchus and returns to Scorpius. (98% illuminated, magnitude -1.5, diameter 16.1", evening sky.)

May 2nd

14:11 UT – The waning crescent Moon is north of Neptune. (Neptune: magnitude 7.9, diameter 2.2". Aquarius, pre-dawn sky.)

May 3rd

Good opportunity to see Earthshine on the waning crescent Moon. (Pre-dawn sky.)

May 5th

02:44 UT – The almost new Moon is south of Uranus. (Uranus: magnitude 5.9, diameter 3.4". Pisces, pre-dawn sky.)

The Eta Aquariid meteor shower peaks. Maximum zenith hourly rate: 65. (Moon: almost new. Aquarius.)

May 6th

03:26 UT – The almost new Moon is south of Venus. (Venus: 99% illuminated, magnitude -3.9, diameter 9.8". Aries, not visible.)

19:29 UT - New Moon (Aries, not visible.)

May 7th

01:07 UT – The just-past new Moon is south of Mercury. (1% illuminated, magnitude 5.1, diameter 11.8". Aries, not visible.)

08:47 UT – The just-past new Moon is south of asteroid 4 Vesta. (Vesta: magnitude 8.1. Taurus, not visible.)

14:41 UT – The just-past new Moon is south of M45, the Pleiades open star cluster. (Taurus, not visible.)

Jupiter decreases its apparent diameter to 40.0". (Magnitude -2.2, Leo, evening sky.)

May 8th

07:45 UT – The waxing crescent Moon is north of Aldebaran. (Taurus, evening sky.)

May 9th

11:12 UT – Mercury begins a transit of the Sun. (0% illuminated, magnitude 6.2, diameter 12.0". Aries.)

15:06 UT – Mercury is at inferior conjunction with the Sun. Distance to Earth: 0.557 AU. (0% illuminated, magnitude 6.2, diameter 12.0". Aries, not visible.)

18:40 UT – Mercury ends its transit of the Sun. (0% illuminated, magnitude 6.2, diameter 12.0". Aries.)

23:12 UT – Jupiter is stationary prior to resuming prograde motion. (Magnitude -2.2, diameter 39.8", Leo, evening sky.)

May 10th

Good opportunity to see Earthshine on the waxing crescent Moon. (Evening sky.)

May 11th

12:09 UT – The waxing crescent Moon is south of Pollux. (Gemini, evening sky.)

17:02 UT – Mercury is at perihelion. Distance to Earth: 0.554 AU. (0% illuminated, magnitude 5.3, diameter 12.0". Aries, not visible.)

May 12th

12:16 UT – The waxing crescent Moon is south of M44, the Praesepe open star cluster. (Cancer, evening sky.)

Asteroid 3 Juno fades to magnitude 10.0. (Virgo, evening sky.)

May 13th

17:02 UT - First Quarter Moon (Leo, evening sky.)

20:38 UT – Mercury is 23' south of Venus. (Mercury: 2% illuminated, magnitude 4.5, diameter 12.0". Venus: 99% illuminated, magnitude -3.9, diameter 9.7". Aries, not visible.)

The Sun leaves Aries and enters Taurus.

May 14th

07:57 UT – The just-past first quarter Moon is south of Regulus (Leo, evening sky.)

May 15th

09:52 UT – The waxing gibbous Moon is south of Jupiter. (Jupiter: magnitude -2.2, diameter 39.1", Leo, evening sky.)

May 17th

Mars brightens to magnitude -2.0. (100% illuminated, diameter 18.0". Scorpius, visible all night.)

May 18th

13:44 UT – The waxing gibbous Moon is north of Spica. (Virgo, evening sky.)

Venus leaves Aries and enters Taurus. (100% illuminated, magnitude -3.9, diameter 9.7", not visible.)

May 19th

16:30 UT - Mercury is at aphelion. Distance to Sun: 0.467 AU. (8% illuminated, magnitude 2.8, diameter 11.3". Aries, not visible.)

May 21st

19:38 UT – The almost full Moon is north of Mars. (Mars: 100% illuminated, magnitude -2.0, diameter 18.3". Scorpius, evening sky.)

21:14 UT - Full Moon (Scorpius, visible all night.)

22:01 UT – Mercury is stationary prior to resuming prograde motion. (12% illuminated, magnitude 2.3, diameter 11.0". Aries, not visible.)

May 22nd

07:58 UT – Mars is at opposition. Distance to Earth: 0.509 AU. (100% illuminated, magnitude -2.1, diameter 18.4". Scorpius, evening sky.)

12:36 UT – The just-past full Moon is north of Antares. (Ophiuchus, all night.)

20:53 UT – The just-past full Moon is north of Saturn. (Saturn: magnitude 0.1, diameter 18.3", Ophiuchus, visible all night.)

May 23rd

01:39 UT – Mars reaches its maximum brightness. (100% illuminated, magnitude -2.1, diameter 18.4". Scorpius, evening sky.)

11:56 UT – Venus is 4.8° south of M45, the Pleiades open star cluster. (Venus: 100% illuminated, magnitude -3.9, diameter 9.6". Taurus, not visible.)

14:25 UT – Asteroid 4 Vesta reaches its minimum brightness. (Magnitude 8.1. Taurus, not visible.)

May 24th

16:59 UT – Asteroid 4 Vesta is in conjunction with the Sun. (Magnitude 8.1. Taurus, not visible.)

Saturn brightens to magnitude 0.0. (Apparent diameter 18.3", Ophiuchus, visible all night.)

May 25th

15:43 UT – The waning gibbous Moon is north of Pluto. (Pluto: magnitude 14.1. Sagittarius, pre-dawn sky.)

Mars fades to magnitude -2.0. (100% illuminated, diameter 18.5". Scorpius, evening sky.)

May 26th

Mercury decreases its apparent diameter to 10.0". (18% illuminated, magnitude 1.6. Aries, pre-dawn sky.)

May 27th

07:28 UT – Asteroid 4 Vesta is at apogee. Distance to Earth: 3.576 AU. (Magnitude 8.1. Taurus, not visible.)

May 29th

01:33 UT – Venus is 3.0° north of asteroid 4 Vesta. (Venus: 100% illuminated, magnitude -3.9, diameter 9.6". Vesta: magnitude 8.1. Taurus, not visible.)

12:12 UT - Last Quarter Moon (Aquarius, pre-dawn.)

19:55 UT – The last quarter Moon is north of Neptune. (Neptune: magnitude 7.9, diameter 2.2". Aquarius, pre-dawn sky.)

Mars leaves Scorpius and returns to Libra. (100% illuminated, magnitude -2.0, diameter 18.6", evening sky.)

May 30th

22:36 UT – Mars is at perigee. Distance to Earth: 0.503 AU. (Mars: 100% illuminated, magnitude -2.0, diameter 18.6". Libra, evening sky.)

May 31ˢᵗ

Mercury brightens to magnitude 1.0. (27% illuminated, diameter 9.2". Aries, pre-dawn sky.)

June

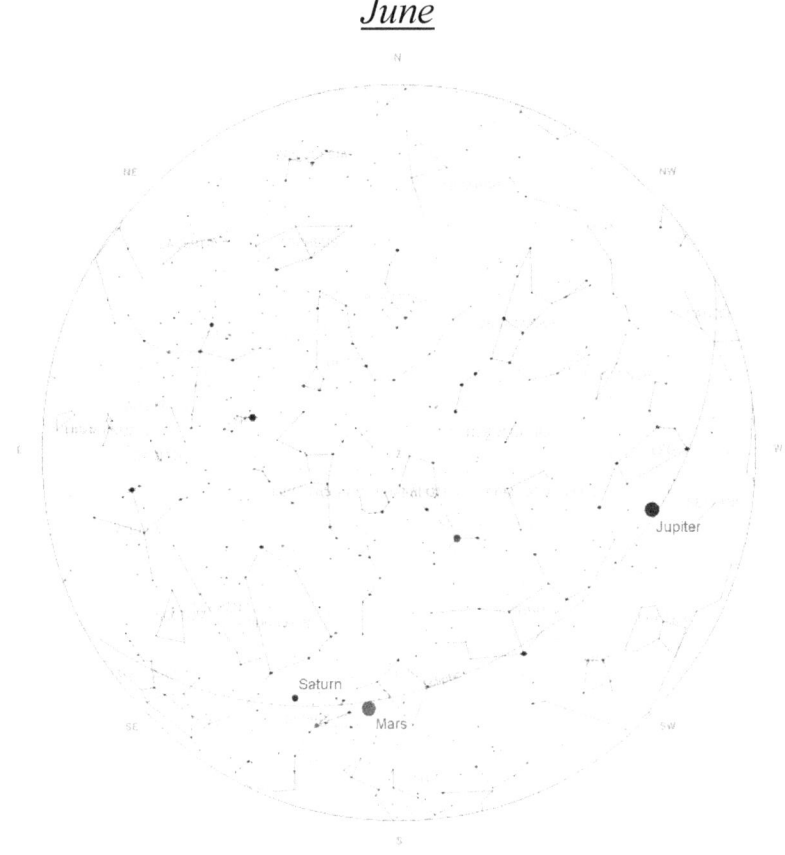

This image depicts the sky at midnight on the 1st but the positions of the stars will be the same at 11:00 p.m. on the 15th and 10:00 p.m. on the 30th. (Daylight Savings Time)

Phases of the Moon

New Moon	First Quarter	Full Moon	Last Quarter
June 5th	June 12th	June 20th	June 27th

June 1st

16:54 UT – The waning crescent Moon is south of Uranus. (Uranus: magnitude 5.9, diameter 3.4". Pisces, pre-dawn sky.)

Jupiter fades to magnitude -2.0 (Diameter 37.2", Leo, evening sky.)

June 2nd

00:31 UT – Venus is 5.3° north of Aldebaran. (Venus: 100% illuminated, magnitude -3.9, diameter 9.6". Taurus, not visible.)

Good opportunity to see Earthshine on the waning crescent Moon. (Pre-dawn sky.)

June 3rd

10:36 UT – Saturn is at perigee. Distance to Earth: 9.015 AU. (Magnitude -0.0, diameter 18.4", Ophiuchus, visible all night.)

11:34 UT – Saturn reaches its maximum brightness. (Magnitude -0.0, diameter 18.4", Ophiuchus, visible all night.)

12:52 UT – Saturn is at opposition. Distance to Earth: 9.015 AU (Magnitude -0.0, diameter 18.4", Ophiuchus, visible all night.)

10:20 UT – The waning crescent Moon is south of Mercury. (Mercury: 34% illuminated, magnitude 0.7, diameter 8.5". Aries, pre-dawn sky.)

June 4th

00:27 UT – The almost new Moon is south of M45, the Pleiades open star cluster. (Taurus, not visible.)

18:12 UT – The almost new Moon is south of asteroid 4 Vesta. (Vesta: magnitude 8.1. Taurus, not visible.)

21:11 UT – The almost new Moon is north of Aldebaran. (Taurus, not visible.)

June 5th

01:46 UT – The almost new Moon is south of Venus. (Venus: 100% illuminated, magnitude -3.9, diameter 9.6". Taurus, not visible.)

02:59 UT - New Moon (Taurus, not visible.)

08:32 UT – Mercury is at Greatest Western Elongation. (38% illuminated, magnitude 0.6, diameter 8.2". Aries, pre-dawn sky.)

Mercury brightens to magnitude 0.5. (37% illuminated, diameter 8.2". Aries, pre-dawn sky.)

June 6th

22:16 UT – Venus is at superior conjunction with the Sun. Distance to Earth: 1.735 AU. (Venus: 100% illuminated, magnitude -3.9, diameter 9.6". Taurus, not visible.)

Mercury leaves Aries and enters Taurus. (39% illuminated, magnitude 0.5, diameter 8.0". Morning sky.)

June 7th

02:05 UT – Venus reaches its maximum brightness. (Venus: 100% illuminated, magnitude -3.9, diameter 9.6". Taurus, not visible.)

03:59 UT – Venus is at apogee. Distance to Earth: 1.735 AU. (Venus: 100% illuminated, magnitude -3.9, diameter 9.6". Taurus, not visible.)

The Arietid meteor shower peaks. Maximum zenith hourly rate: 54. (Moon: waxing crescent. Aries.)

June 8th

00:30 UT – The waxing crescent Moon is south of Pollux. (Gemini, evening sky.)

Good opportunity to see Earthshine on the waxing crescent Moon. (Evening sky.)

June 9th

00:21 UT – The waxing crescent Moon is south of M44, the Praesepe open star cluster. (Cancer, evening sky.)

11:28 UT – Asteroid 4 Vesta is 2.4° north of Aldebaran. (Vesta: magnitude 8.1. Taurus, not visible.)

June 10th

14:20 UT – The waxing crescent Moon is south of Regulus (Leo, evening sky.)

June 11th

10:12 UT – Mercury is 50% illuminated. (Magnitude 0.1, diameter 7.2". Taurus, pre-dawn sky.)

16:51 UT – Mercury is 7.2° south of M45, the Pleiades open star cluster. (Mercury: 34% illuminated, magnitude 0.1, diameter 7.1". Taurus, pre-dawn sky.)

22:08 UT – The almost first quarter Moon is south of Jupiter. (Jupiter: magnitude -2.0, diameter 36.1", Leo, evening sky.)

June 12th

08:09 UT - First Quarter Moon (Leo, evening sky.)

18:00 UT - Mercury brightens to magnitude 0.0. (53% illuminated, diameter 6.9". Taurus, pre-dawn sky.)

June 14th

22:40 UT – The waxing gibbous Moon is north of Spica. (Virgo, evening sky.)

June 15th

04:59 UT – Neptune is stationary prior to beginning retrograde motion. (Neptune: magnitude 7.9, diameter 2.3". Aquarius, pre-dawn sky.)

Asteroid 2 Pallas brightens to magnitude 9.5. (Pegasus, pre-dawn sky.)

June 17th

11:10 UT – The waxing gibbous Moon is north of Mars. (Mars: 97% illuminated, magnitude -1.7, diameter 17.7". Libra, evening sky.)

Venus leaves Taurus and enters Gemini. (100% illuminated, magnitude -3.9, diameter 9.6", not visible.)

June 18th

01:10 UT – Asteroid 2 Pallas is stationary prior to beginning retrograde motion. (Magnitude 9.5. Pegasus, pre-dawn sky.)

17:06 UT – The waxing gibbous Moon is north of Antares. (Ophiuchus, evening sky.)

Mercury brightens to magnitude -0.5. (65% illuminated, diameter 6.3". Taurus, pre-dawn sky.)

June 19th

02:24 UT – The almost full Moon is north of Saturn. (Saturn: magnitude 0.1, diameter 18.3", Ophiuchus, visible all night.)

17:37 UT – Mercury is 3.9° north of Aldebaran. (Mercury: 69% illuminated, magnitude -0.6, diameter 6.1". Taurus, pre-dawn sky.)

June 20th

11:02 UT - Full Moon. This is the southernmost full moon of the year. (Sagittarius, visible all night.)

22:34 UT - Northern Solstice. Summer begins in the northern hemisphere, winter begins in the southern hemisphere.

June 21st

18:50 UT – The just-past full Moon is north of Pluto. (Pluto: magnitude 14.1. Sagittarius, visible all night.)

The Sun leaves Taurus and enters Gemini.

Jupiter decreases its apparent diameter to 35.0". (Magnitude -1.9, Leo, evening sky.)

Asteroid 3 Juno fades to magnitude 10.5. (Virgo, evening sky.)

June 23rd

01:42 UT – Mercury is 2.0° north of asteroid 4 Vesta. (Mercury: 78% illuminated, magnitude -0.8, diameter 5.7". Taurus, pre-dawn sky. Vesta: magnitude 8.2. Taurus, not visible.)

Uranus brightens to magnitude 5.8. (Apparent diameter 3.4". Pisces, pre-dawn sky.)

June 24th

Mercury brightens to magnitude -1.0. (80% illuminated, diameter 5.7". Taurus, pre-dawn sky.)

Mars fades to magnitude -1.5. (95% illuminated, diameter 17.1". Libra, evening sky.)

Uranus increases its apparent diameter to 3.5". (Magnitude 5.8, Pisces, pre-dawn sky.)

June 25th

04:51 UT – Asteroid 3 Juno is stationary prior to resuming prograde motion. (Magnitude 10.5. Virgo, evening sky.)

23:48 UT – The waning gibbous Moon is north of Neptune. (Neptune: magnitude 7.9, diameter 2.3". Aquarius, pre-dawn sky.)

June 27th

18:18 UT - Last Quarter Moon (Cetus, pre-dawn sky.)

June 28th

22:29 UT – The just-past last quarter Moon is south of Uranus. (Uranus: magnitude 5.8, diameter 3.5". Pisces, pre-dawn sky.)

June 29th

Mercury brightens to magnitude -1.5. (92% illuminated, diameter 5.3". Taurus, not visible.)

June 30th

09:52 UT – Mars is stationary prior to resuming prograde motion. (Mars: 93% illuminated, magnitude -1.4, diameter 16.4". Libra, evening sky.)

Mercury leaves Taurus and enters Gemini. (94% illuminated, magnitude -1.5, diameter 5.2", not visible.)

Dwarf planet Ceres brightens to magnitude 8.5. (Cetus, pre-dawn sky.)

July

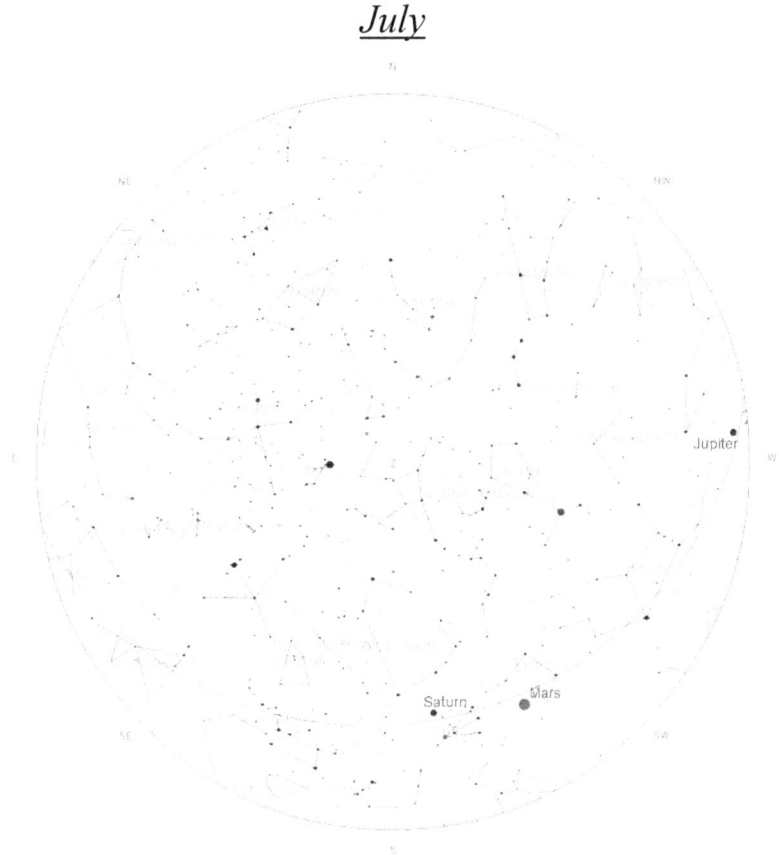

This image depicts the sky at midnight on the 1st but the positions of the stars will be the same at 11:00 p.m. on the 15th and 10:00 p.m. on the 31st. (Daylight Savings Time)

Phases of the Moon

New Moon	First Quarter	Full Moon	Last Quarter
July 4th	July 12th	July 19th	July 26th

July 1st

08:38 UT – The waning crescent Moon is south of M45, the Pleiades open cluster. (Taurus, pre-dawn sky.)

Good opportunity to see Earthshine on the waning crescent Moon. (Pre-dawn sky.)

July 2nd

03:26 UT – The waning crescent Moon is north of Aldebaran. (Taurus, pre-dawn sky.)

16:08 UT - Mercury is at perihelion. Distance to Sun: 0.308 AU. (97% illuminated, magnitude -1.8, diameter 5.1". Gemini, not visible.)

22:19 UT – The waning crescent Moon is south of asteroid 4 Vesta. (Vesta: magnitude 8.2. Taurus, not visible.)

July 3rd

11:48 UT – Pluto reaches its maximum brightness. (Pluto: magnitude 14.1. Sagittarius, visible all night.)

July 4th

04:38 UT – The almost new Moon is south of Mercury (Mercury: 99% illuminated, diameter 5.1", magnitude -2.0, Gemini, not visible.)

11:01 UT - New Moon (Gemini, not visible.)

15:45 UT – Spring begins in the southern hemisphere on Mars. Dust storm season begins. (92% illuminated, magnitude -1.3, diameter 15.9". Libra, evening sky.)

18:00 UT – The Earth is at aphelion. Distance to Sun: 1.017 AU.

Mercury brightens to magnitude -2.0. (99% illuminated, diameter 5.1". Gemini, not visible.)

July 5th

03:02 UT – The just-past new Moon is south of Venus. (Venus: 99% illuminated, magnitude -3.9, diameter 9.7". Gemini, not visible.)

07:17 UT – The just-past new Moon is south of Pollux. (Gemini, evening sky, not visible.)

13:28 UT – Pluto is at perigee. Distance to Earth: 32.114 AU. (Pluto: magnitude 14.1. Sagittarius, visible all night.)

Mercury decreases its apparent diameter to 5.0". (100% illuminated, magnitude -2.1. Gemini, not visible.)

Saturn decreases its apparent diameter to 18.0". (Magnitude 0.2, Ophiuchus, evening sky.)

July 6th

06:45 UT – The waxing crescent Moon is south of M44, the Praesepe open star cluster. (Cancer, evening sky.)

20:39 UT – Mercury reaches its maximum brightness. (100% illuminated, magnitude -2.2, diameter 5.0". Gemini, not visible.)

July 7th

03:11 UT – Mercury is at superior conjunction with the Sun. Distance to Earth: 1.329 AU. (100% illuminated, magnitude -2.2, diameter 5.0". Gemini, not visible.)

13:18 UT – Pluto is at opposition. Distance to Earth: 32.115 AU. (Pluto: magnitude 14.1. Sagittarius, visible all night.)

13:34 UT – Venus is 5.7° south of Pollux. (Venus: 99% illuminated, magnitude -3.9, diameter 9.8". Gemini, not visible.)

July 8th

01:38 UT – The waxing crescent Moon is south of Regulus. (Leo, evening sky.)

Good opportunity to see Earthshine on the waxing crescent Moon. (Evening sky.)

July 9th

00:32 UT – Mercury is at apogee. Distance to Earth: 1.332 AU. (99% illuminated, magnitude -2.0, diameter 5.0". Gemini, not visible.)

09:29 UT – The waxing crescent Moon is south of Jupiter. (Jupiter: magnitude -1.8, diameter 33.5", Leo, evening sky.)

Neptune brightens to magnitude 7.8. (Diameter 2.3". Aquarius, pre-dawn sky.)

July 10th

21:59 UT – Venus is at perihelion. Distance to Sun: 0.719 AU. (Venus: 99% illuminated, magnitude -3.9, diameter 9.8". Cancer, not visible.)

Venus leaves Gemini and enters Cancer. (99% illuminated, magnitude -3.9, diameter 9.8", not visible.)

July 11th

04:23 UT – Mercury is 5.1° south of Pollux. (98% illuminated, magnitude -1.7, diameter 5.0". Gemini, not visible.)

July 12th

00:51 UT - First Quarter Moon (Virgo, evening sky.)

05:49 UT – The first quarter Moon is north of Spica (Virgo, evening sky.)

Mercury leaves Gemini and enters Cancer. (97% illuminated, magnitude -1.6, diameter 5.0", not visible.)

Mercury fades to magnitude -1.5. (97% illuminated, diameter 5.0". Cancer, not visible.)

Mars decreases its apparent diameter to 15.0". (91% illuminated, magnitude -1.2. Libra, evening sky.)

July 14th

18:13 UT – The waxing gibbous Moon is north of Mars. (Mars: 90% illuminated, magnitude -1.1, diameter 14.8". Libra, evening sky.)

July 16th

04:22 UT – The waxing gibbous Moon is north of Antares. (Ophiuchus, evening sky.)

07:03 UT – The waxing gibbous Moon is north of Saturn. (Saturn: magnitude 0.3, diameter 17.8", Ophiuchus, evening sky.)

16:36 UT – Asteroid Vesta is at aphelion. Distance to Sun: 2.571 AU. (Magnitude 8.2. Taurus, pre-dawn sky.)

18:31 UT – Mercury is 31' north of Venus. (Mercury: 93% illuminated, magnitude -1.0, diameter 5.1". Venus: 98% illuminated, magnitude -3.9, diameter 9.9". Cancer, not visible.)

Mercury fades to magnitude -1.0. (93% illuminated, diameter 5.1". Cancer, not visible.)

July 17th

15:33 UT – Mercury is 36' north of M44, the Praesepe open star cluster. (91% illuminated, magnitude -1.0, diameter 5.2". Cancer, not visible.)

Mars fades to magnitude -1.0. (90% illuminated, diameter 14.5". Libra, evening sky.)

July 18th

03:43 UT – Venus is 6' north of M44, the Praesepe open star cluster. (Venus: 98% illuminated, magnitude -3.9, diameter 9.9". Cancer, not visible.)

July 19th

04:05 UT – The almost full Moon is north of Pluto. (Pluto: magnitude 14.1. Sagittarius, visible all night.)

22:56 UT - Full Moon (Sagittarius, visible all night.)

July 20th

The Sun leaves Gemini and enters Cancer.

Asteroid 4 Vesta leaves Taurus and enters Orion. (Magnitude 8.2, pre-dawn sky.)

July 22nd

Venus increases its apparent diameter to 10.0". (98% illuminated, magnitude -3.9. Cancer, not visible.)

July 23rd

08:13 UT – The waning gibbous Moon is north of Neptune. (Neptune: magnitude 7.8, diameter 2.3". Aquarius, pre-dawn sky.)

Mercury leaves Cancer and enters Leo. (84% illuminated, magnitude -0.6, diameter 5.4", evening sky.)

Mercury fades to magnitude -0.5. (84% illuminated, diameter 5.4". Leo, evening sky.)

July 26th

04:54 UT – The almost last quarter Moon is south of Uranus. (Uranus: magnitude 5.8, diameter 3.5". Pisces, pre-dawn sky.)

22:59 UT - Last Quarter. This is the largest last quarter Moon of the year, the largest of the past ten years and the largest for the next forty years. Apparent diameter: 32.308' (Cetus, pre-dawn sky.)

Venus leaves Cancer and enters Leo. (97% illuminated, magnitude -3.9, diameter 10.0", evening sky.)

July 28th

16:58 UT – The just-past last quarter Moon is south of M45, the Pleiades open star cluster. (Taurus, pre-dawn sky.)

Asteroid 2 Pallas brightens to magnitude 9.0. (Pegasus, evening sky.)

July 29th

13:26 UT – The waning crescent Moon is north of Aldebaran. (Taurus, pre-dawn sky.)

23:34 UT – Uranus is stationary prior to beginning retrograde motion. (Uranus: magnitude 5.8, diameter 3.6". Pisces, pre-dawn sky.)

The Southern Delta Aquariid meteor shower peaks. Maximum zenith hourly rate: 16. (Moon: waning crescent. Aquarius.)

The Beta Cassiopeid meteor shower peaks. Maximum zenith hourly rate: 10. (Moon: waning crescent. Cassiopeia.)

July 30th

14:36 UT – Mercury is 20' north of Regulus. (Mercury: 74% illuminated, magnitude -0.2, diameter 5.8". Leo, evening sky.)

July 31st

Good opportunity to see Earthshine on the waning crescent Moon. (Pre-dawn sky.)

August

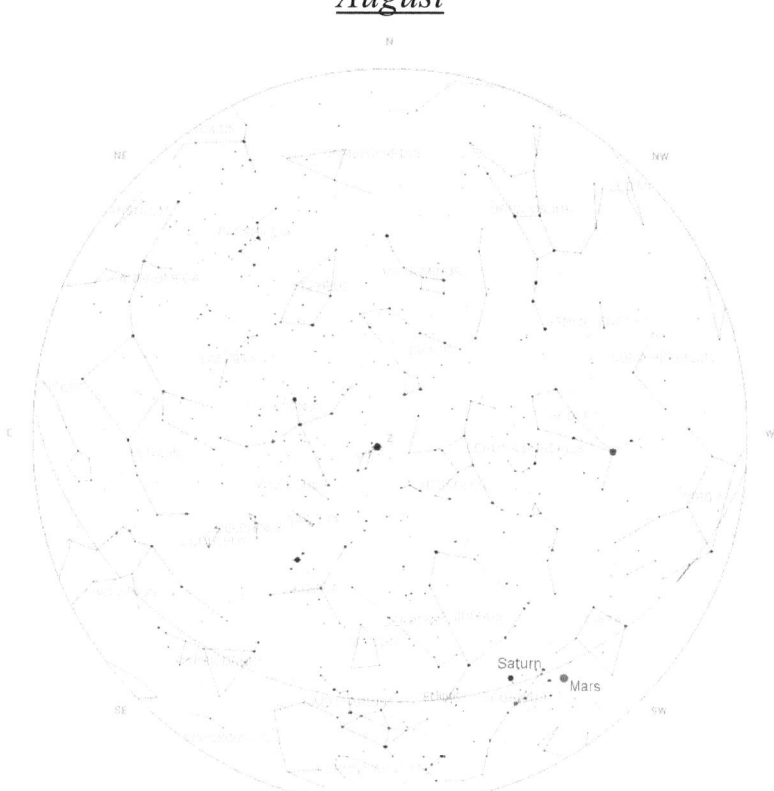

This image depicts the sky at 11:00 p.m. on the 1st but the positions of the stars will be the same at 10:00 p.m. on the 15th and 9:00 p.m. on the 31st. (Daylight Savings Time)

Phases of the Moon

New Moon	First Quarter	Full Moon	Last Quarter
August 2nd	August 10th	August 18th	August 25th

August 1st

19:13 UT – The almost new Moon is south of Pollux. (Gemini, not visible.)

Mars leaves Libra and returns to Scorpius. (87% illuminated, magnitude -0.8, diameter 13.0", evening sky.)

August 2nd

18:51 UT – The almost new Moon is south of M44, the Praesepe open star cluster. (Cancer, not visible.)

20:44 UT - New Moon (Cancer, not visible.)

August 4th

05:43 UT – The waxing crescent Moon is south of Venus. (Venus: 96% illuminated, magnitude -3.9, diameter 10.2". Leo, evening sky.)

07:35 UT – The waxing crescent Moon is south of Regulus. (Leo, evening sky.)

11:00 UT - Mercury fades to magnitude 0.0. (68% illuminated, diameter 6.2". Leo, evening sky.)

August 5th

00:22 UT – The waxing crescent Moon is south of Mercury. (Mercury: 67% illuminated, magnitude 0.0, diameter 6.2". Leo, evening sky.)

05:22 UT – Venus is 1.1° north of Regulus. (Venus: 96% illuminated, magnitude -3.9, diameter 10.2". Leo, evening sky.)

August 6th

04:13 UT – The waxing crescent Moon is south of Jupiter. (Jupiter: magnitude -1.7, diameter 31.7", Leo, evening sky.)

Asteroid 3 Juno fades to magnitude 11.0. (Virgo, evening sky.)

Good opportunity to see Earthshine on the waxing crescent Moon. (Evening sky.)

August 7th

Jupiter leaves Leo and enters Virgo. (Magnitude -1.7, diameter 31.7", evening sky.)

Asteroid 4 Vesta leaves Orion and enters Gemini. (Magnitude 8.1, pre-dawn sky.)

August 8th

11:13 UT – The waxing crescent Moon is north of Spica. (Virgo, evening sky.)

August 10th

18:21 UT - First Quarter Moon. This is the smallest first quarter moon of the year. (Libra, evening sky.)

The Sun leaves Cancer and enters Leo.

August 12th

00:43 UT – The waxing gibbous Moon is north of Mars. (Mars: 86% illuminated, magnitude -0.6, diameter 11.9". Scorpius, evening sky.)

09:57 UT – The waxing gibbous Moon is north of Antares. (Ophiuchus, evening sky.)

11:25 UT – The waxing gibbous Moon is north of Saturn. (Saturn: magnitude 0.4, diameter 17.1", Ophiuchus, evening sky.)

The Perseid meteor shower peaks. Maximum zenith hourly rate: 100. (Moon: waxing gibbous. Perseus.)

August 13th

04:03 UT – Asteroid 2 Pallas is at opposition. (Magnitude 8.9. Pegasus, visible all night.)

14:12 UT – Saturn is stationary prior to resuming prograde motion. (Magnitude 0.4, diameter 17.1", Ophiuchus, evening sky.)

August 14th

Mars fades to magnitude -0.5. (86% illuminated, diameter 11.8". Scorpius, evening sky.)

Dwarf planet Ceres brightens to magnitude 8.0. (Cetus, pre-dawn sky.)

August 15th

10:21 UT – The waxing gibbous Moon is north of Pluto. (Pluto: magnitude 14.2. Sagittarius, evening sky.)

16:45 UT – Mercury is at aphelion. Distance to Sun: 0.467 AU. (53% illuminated, magnitude 0.3, diameter 7.2". Leo, evening sky.)

Saturn decreases its apparent diameter to 17.0". (Magnitude 0.4, Ophiuchus, evening sky.)

August 16th

22:08 UT – Mercury reaches Greatest Eastern Elongation. (51% illuminated, magnitude 0.3, diameter 7.4". Leo, evening sky.)

August 17th

10:54 UT – Mercury is 50% illuminated. (Magnitude 0.4, diameter 7.4". Leo, evening sky.)

August 18th

09:26 UT - Full Moon (Capricornus, visible all night.)

Uranus brightens to magnitude 5.7. (Apparent diameter 3.6". Pisces, pre-dawn sky.)

August 19th

13:20 UT – The just-past full Moon is north of Neptune. (Neptune: magnitude 7.8, diameter 2.3". Aquarius, visible all night.)

August 20th

Mercury leaves Leo and enters Virgo. (95% illuminated, magnitude 0.4, diameter 7.8", evening sky.)

Mercury fades to magnitude 0.5. (95% illuminated, diameter 7.8". Virgo, evening sky.)

Saturn fades to magnitude 0.5. (Apparent diameter 16.9", Ophiuchus, evening sky.)

August 21st

Mars leaves Scorpius and returns to Ophiuchus. (86% illuminated, magnitude -0.4, diameter 11.2", evening sky.)

Asteroid 2 Pallas leaves Pegasus and returns to Equuleus. (Magnitude 8.9, evening sky.)

August 22nd

12:27 UT – The waning gibbous Moon is south of Uranus. (Uranus: magnitude 5.7, diameter 3.6". Pisces, pre-dawn sky.)

August 23rd

18:07 UT – Mars is 1.8° north of Antares. (Mars: 85% illuminated, magnitude -0.4, diameter 11.1". Ophiuchus, evening sky.)

August 24th

05:30 UT – Asteroid 2 Pallas reaches its maximum brightness. (Magnitude 8.9. Equuleus, visible all night.)

14:23 UT – Asteroid 2 Pallas is at perigee. Distance to Earth: 2.398 AU. (Magnitude 8.9. Equuleus, visible all night.)

20:15 UT – The almost last quarter Moon is south of M45, the Pleiades open star cluster. (Taurus, pre-dawn sky.)

Venus leaves Leo and enters Virgo. (93% illuminated, magnitude -3.9, diameter 10.7", evening sky.)

August 25th

03:41 UT - Last Quarter Moon (Taurus, pre-dawn sky.)

16:57 UT – The last quarter Moon is south of Aldebaran. An occultation will be visible from some parts of the world. (Taurus, pre-dawn sky.)

18:46 UT – Mars is 4.4° south of Saturn. (Mars: 85% illuminated, magnitude -0.4, diameter 10.9". Saturn: magnitude 0.5, diameter 16.8". Ophiuchus, evening sky.)

August 26th

Mars leaves Ophiuchus and returns to Scorpius. (85% illuminated, magnitude -0.4, diameter 10.9", evening sky.)

August 27th

05:44 UT – Mercury is 5.3° south of Venus. (Mercury: 32% illuminated, magnitude 0.8, diameter 8.8". Venus: 93% illuminated, magnitude -3.9, diameter 10.8". Virgo, evening sky.)

22:46 UT – Venus is 4' north of Jupiter. (Venus: 92% illuminated, magnitude -3.9, diameter 10.8". Jupiter: magnitude -1.7, diameter 30.9". Virgo, evening sky.)

Asteroid 4 Vesta brightens to magnitude 8.0. (Gemini, pre-dawn sky.)

August 28th

23:46 UT – The waning crescent Moon is south of Pollux. (Gemini, pre-dawn sky.)

Mercury fades to magnitude 1.0. (30% illuminated, diameter 8.9". Virgo, evening sky.)

August 29th

23:59 UT – The waning crescent Moon is south of M44, the Praesepe open star cluster. (Cancer, pre-dawn sky.)

Good opportunity to see Earthshine on the waning crescent Moon. (Pre-dawn sky.)

August 30th

01:41 UT – Mercury is stationary prior to beginning retrograde motion. (26% illuminated, magnitude 1.1, diameter 9.3". Virgo, evening sky.)

August 31st

18:59 UT – The almost new Moon is south of Regulus. (Leo, not visible.)

September

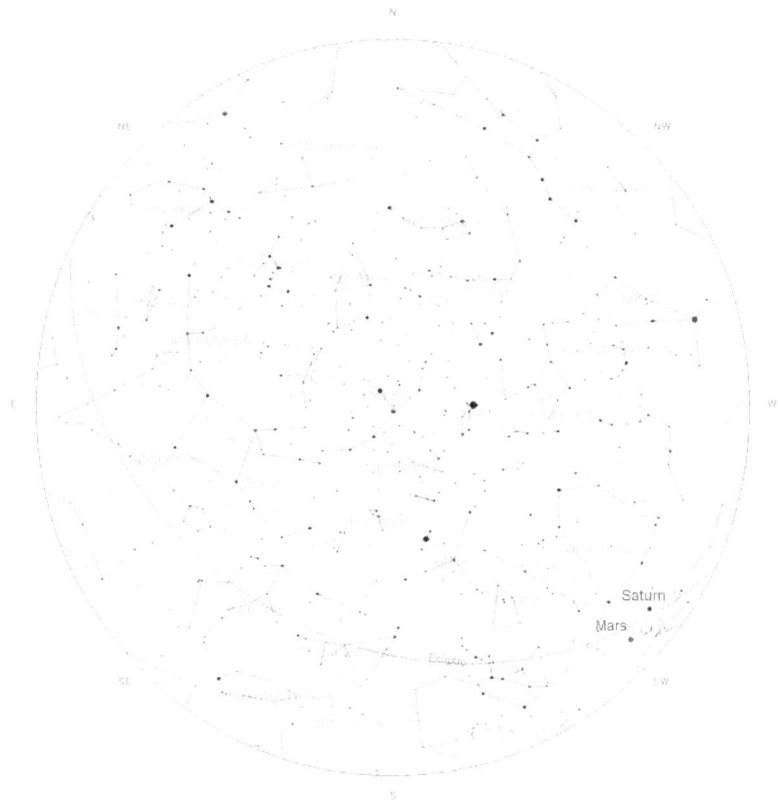

This image depicts the sky at 11:00 p.m. on the 1st but the positions of the stars will be the same at 10:00 p.m. on the 15th and 9:00 p.m. on the 30th. (Daylight Savings Time)

Phases of the Moon

New Moon	First Quarter	Full Moon	Last Quarter
September 1st	September 9th	September 16th	September 23rd

September 1ˢᵗ

06:14 UT – First location to see a partial solar eclipse begin.

07:18 UT – First location to see the annular solar eclipse begin.

09:02 UT – Annular solar eclipse maximum. Visible from most of Africa, Antarctic, south Asia, the Atlantic, eastern Australia and the Indian Ocean.

09:03 UT - New Moon (Leo, not visible.)

10:56 UT – Last location to see the annular solar eclipse end.

23:28 UT – Neptune is at perigee. Distance to Earth: 28.945 AU. (Neptune: magnitude 7.8, diameter 2.3". Aquarius, visible all night.)

September 2ⁿᵈ

01:00 UT – Neptune reaches its maximum brightness. (Neptune: magnitude 7.8, diameter 2.3". Aquarius, visible all night.)

01:00 UT – Dwarf planet Ceres is stationary prior to beginning retrograde motion. (Magnitude 7.8. Cetus, pre-dawn sky.)

20:16 UT – The just-past new Moon is north of Mercury. (Mercury: 17% illuminated, magnitude 1.7, diameter 9.8". Virgo, not visible.)

23:58 UT – The just-past new Moon is north of Jupiter. (Jupiter: magnitude -1.7, diameter 30.8", Virgo, evening sky.)

Mars leaves Scorpius and returns to Ophiuchus. (85% illuminated, magnitude -0.3, diameter 10.4", evening sky.)

September 3ʳᵈ

03:00 UT – Neptune is at opposition. Distance to Earth: 28.946 AU. (Neptune: magnitude 7.8, diameter 2.3". Aquarius, visible all night.)

09:33 UT – The waxing crescent Moon is north of Venus. (Venus: 91% illuminated, magnitude -3.9, diameter 11.0". Virgo, evening sky.)

Mercury increases its apparent diameter to 10.0". (17% illuminated, magnitude 1.7. Virgo, not visible.)

September 4th

22:44 UT – The waxing crescent Moon is north of Spica. (Virgo, evening sky.)

September 5th

Asteroid 3 Juno leaves Virgo and returns to Libra (Magnitude 11.2, evening sky.)

Good opportunity to see Earthshine on the waxing crescent Moon. (Evening sky.)

September 7th

Mercury leaves Virgo and returns to Leo. (8% illuminated, magnitude 2.8, diameter 10.4", not visible.)

Mars decreases its apparent diameter to 10.0". (85% illuminated, magnitude -0.2. Ophiuchus, evening sky.)

September 8th

19:28 UT – The almost first quarter Moon is north of Antares. (Ophiuchus, evening sky.)

23:52 UT – The almost first quarter Moon is north of Saturn. (Saturn: magnitude 0.5, diameter 16.4", Ophiuchus, evening sky.)

September 9th

11:49 UT - First Quarter Moon (Ophiuchus, evening sky.)

12:52 UT – The first quarter Moon is north of Mars. (Mars: 85% illuminated, magnitude -0.2, diameter 9.9". Ophiuchus, evening sky.)

Asteroid 2 Pallas fades to magnitude 9.0. (Equuleus, evening sky.)

September 10th

09:51 UT – Mercury is at perigee. Distance to Earth: 0.635 AU. (2% illuminated, magnitude 4.1, diameter 10.5". Leo, not visible.)

September 11th

17:34 UT – The waxing gibbous Moon is north of Pluto. (Pluto: magnitude 14.2. Sagittarius, evening sky.)

September 13th

00:34 UT – Mercury is at inferior conjunction with the Sun. Distance to Earth: 0.642 AU. (1% illuminated, magnitude 4.9, diameter 10.4". Leo, not visible.)

September 15th

19:58 UT – The almost full Moon is north of Neptune. (Neptune: magnitude 7.8, diameter 2.3". Aquarius, visible all night.)

Mercury decreases its apparent diameter to 10.0". (1% illuminated, magnitude 4.4. Leo, not visible.)

September 16th

16:57 UT – Penumbral lunar eclipse begins.

18:55 UT – Penumbral lunar eclipse maximum. Visible from Africa, Antarctica, the Arctic, most of Asia, the Atlantic, Australia, Europe, the Indian and Pacific oceans and western South America.

19:05 UT - Full Moon (Pisces, visible all night.)

20:53 UT – Penumbral lunar eclipse ends.

The Sun leaves Leo and enters Virgo.

September 17th

19:10 UT – Venus is 2.7° north of Spica. (Venus: 88% illuminated, magnitude -3.9, diameter 11.5". Virgo, evening sky.)

September 18th

16:50 UT – The waning gibbous Moon is south of Uranus. (Uranus: magnitude 5.7, diameter 3.7". Pisces, pre-dawn sky.)

Dwarf planet Ceres brightens to magnitude 7.5. (Cetus, pre-dawn sky.)

September 19th

04:36 UT – Asteroid 3 Juno is at aphelion. Distance to Sun: 3.351 AU. (Magnitude 10.5. Virgo, evening sky.)

September 21st

02:18 UT – The waning gibbous Moon is south of M45, the Pleiades open star cluster. (Taurus, pre-dawn sky.)

10:47 UT – Mercury is stationary prior to resuming prograde motion. (17% illuminated, magnitude 1.3, diameter 8.8". Leo, not visible.)

20:43 UT – The waning gibbous Moon is south of Aldebaran. An occultation may be visible from some parts of the world. (Taurus, pre-dawn sky.)

Mars leaves Ophiuchus and enters Sagittarius. (85% illuminated, magnitude -0.0, diameter 9.3", evening sky.)

Saturn decreases its apparent diameter to 16.0". (Magnitude 0.6, Ophiuchus, evening sky.)

September 22nd

14:21 UT - Autumnal Equinox. Autumn begins in the northern hemisphere, spring begins in the southern hemisphere.

Mercury brightens to magnitude 1.0. (19% illuminated, diameter 8.6". Leo, not visible.)

September 23rd

09:56 UT - Last Quarter Moon. This is the northernmost last quarter moon of the year. (Orion, pre-dawn sky.)

Mercury brightens to magnitude 0.5. (23% illuminated, diameter 8.4". Leo, not visible.)

September 24th

23:00 UT - Mars fades to magnitude 0.0. (85% illuminated, diameter 9.1". Sagittarius, evening sky.)

September 25th

04:27 UT – The waning crescent Moon is south of Pollux. (Gemini, pre-dawn sky.)

08:48 UT – Jupiter is at apogee. Distance to Earth: 6.454 AU. (Magnitude -1.7, diameter 30.5", Virgo, not visible.)

19:00 UT – Pluto is stationary prior to resuming prograde motion. (Magnitude 14.2. Sagittarius, evening sky.)

September 26th

04:29 UT – The waning crescent Moon is south of M44, the Praesepe open star cluster. (Cancer, pre-dawn sky.)

05:00 UT - Mercury brightens to magnitude 0.0. (38% illuminated, diameter 7.6". Leo, pre-dawn sky.)

22:56 UT – Jupiter is in conjunction with the Sun. Distance to Earth: 6.454 AU. (Jupiter: magnitude -1.7, diameter 30.5", Virgo, not visible.)

Asteroid 2 Pallas leaves Equuleus and enters Aquarius. (Magnitude 9.1, evening sky.)

September 27th

23:15 UT – The waning crescent Moon is south of Regulus. (Leo, pre-dawn sky.)

September 28th

16:23 UT – Mercury is at perihelion. Distance to Sun: 0.308 AU. (50% illuminated, magnitude -0.4, diameter 7.0". Leo, pre-dawn sky.)

17:42 UT - Mercury is 50% illuminated. (Magnitude -0.4, diameter 7.0". Leo, pre-dawn sky.)

20:18 UT – Mercury reaches Greatest Western Elongation. (50% illuminated, magnitude -0.4, diameter 7.0". Leo, pre-dawn sky.)

Mercury brightens to magnitude -0.5. (47% illuminated, diameter 7.2". Leo, pre-dawn sky.)

Good opportunity to see Earthshine on the waning crescent Moon. (Pre-dawn sky.)

September 29th

11:37 UT – The waning crescent Moon is south of Mercury. (Mercury: 53% illuminated, magnitude -0.5, diameter 6.9". Leo, pre-dawn sky.)

September 30th

19:07 UT – The almost new Moon is north of Jupiter. (Jupiter: magnitude -1.7, diameter 30.5", Virgo, not visible.)

Venus leaves Virgo and enters Libra. (86% illuminated, magnitude -3.9, diameter 12.1", evening sky.)

October

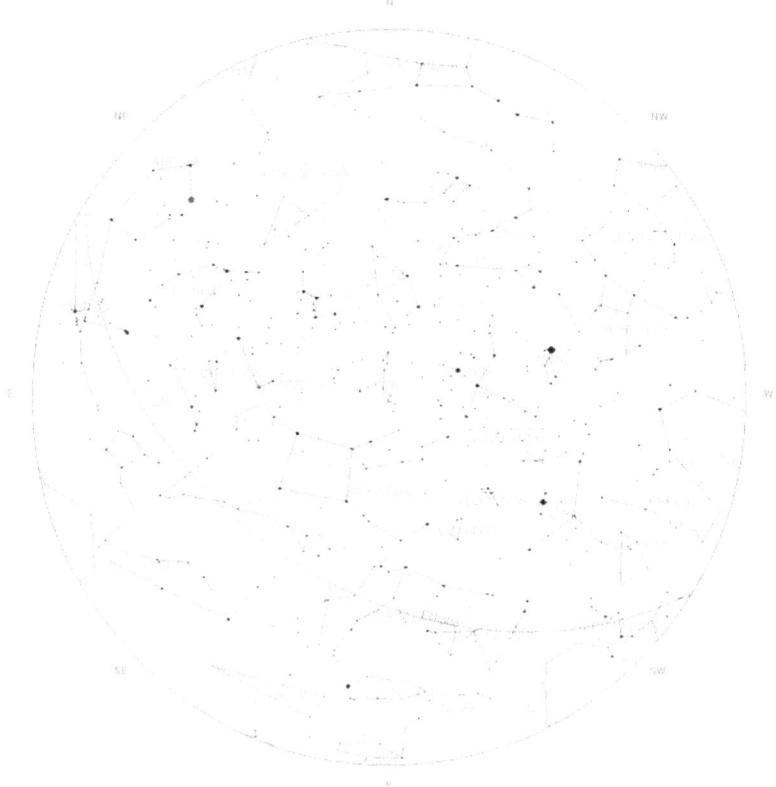

This image depicts the sky at 11:00 p.m. on the 1st but the positions of the stars will be the same at 10:00 p.m. on the 15th and 9:00 p.m. on the 31st. (Daylight Savings Time)

Phases of the Moon

New Moon	First Quarter	Full Moon	Last Quarter	New Moon
October 1st	October 9th	October 16th	October 22nd	October 30th

October 1st

00:11 UT - The first of two New Moons for October. The second occurs on the 30th. (Virgo, not visible.)

October 2nd

03:16 UT – The just-past new Moon is north of Spica. (Virgo, not visible.)

Mercury leaves Leo and returns to Virgo. (64% illuminated, magnitude -0.8, diameter 6.4", pre-dawn sky.)

October 3rd

20:37 UT – The waxing crescent Moon is north of Venus. (Venus: 85% illuminated, magnitude -3.9, diameter 12.3". Libra, evening sky.)

October 4th

Good opportunity to see Earthshine on the waxing crescent Moon. (Evening sky.)

October 5th

Mercury brightens to magnitude -1.0. (75% illuminated, diameter 5.9". Virgo, pre-dawn sky.)

October 6th

02:52 UT – The waxing crescent Moon is north of Antares. (Ophiuchus, evening sky.)

07:29 UT – The waxing crescent Moon is north of Saturn. (Saturn: magnitude 0.6, diameter 15.7", Ophiuchus, evening sky.)

Venus brightens to magnitude -4.0. (84% illuminated, diameter 12.4". Libra, evening sky.)

October 7th

06:31 UT – Asteroid 2 Pallas is stationary prior to resuming prograde motion. (Magnitude 9.2. Aquarius, evening sky.)

October 8th

11:19 UT – The almost first quarter Moon is north of Mars. (Mars: 85% illuminated, magnitude 0.1, diameter 8.4". Sagittarius, evening sky.)

The Draconid meteor shower peaks. Maximum zenith hourly rate: Variable. (Moon: almost first quarter. Draco.)

October 9th

04:18 UT – The almost first quarter Moon is north of Pluto. (Pluto: magnitude 14.2. Sagittarius, evening sky.)

04:33 UT - First Quarter Moon. This is the southernmost first quarter moon of the year. (Sagittarius, evening sky.)

Asteroid 4 Vesta leaves Gemini and enters Cancer. (Magnitude 7.7, pre-dawn sky.)

October 11th

05:13 UT – Mercury is 48' north of Jupiter. (Mercury: 90% illuminated, magnitude -1.1, diameter 5.3". Jupiter: magnitude -1.7, diameter 30.6". Virgo, not visible.)

October 13th

08:00 UT – The waxing gibbous Moon is north of Neptune. (Neptune: magnitude 7.8, diameter 2.3". Aquarius, evening sky.)

October 14th

20:46 UT – Uranus is at perigee. Distance to Earth: 18.951 AU. (Uranus: magnitude 5.7, diameter 3.7". Pisces, visible all night.)

October 15th

04:00 UT – Uranus reaches its maximum brightness. (Uranus: magnitude 5.7, diameter 3.7". Pisces, visible all night.)

15:43 UT – Uranus is at opposition. Distance to Earth: 18.951 AU. (Uranus: magnitude 5.7, diameter 3.7". Pisces, visible all night.)

Mercury decreases its apparent diameter to 5.0". (95% illuminated, magnitude -1.2. Virgo, not visible.)

October 16th

03:43 UT – The almost full Moon is south of Uranus. (Uranus: magnitude 5.7, diameter 3.7". Pisces, visible all night.)

04:23 UT - Full Moon (Pisces, visible all night.)

October 17th

Venus leaves Libra and enters Scorpius. (82% illuminated, magnitude -4.0, diameter 13.0", evening sky.)

October 18th

12:55 UT - The waning gibbous Moon is south of M45, the Pleiades open star cluster. (Taurus, pre-dawn sky.)

15:32 UT – Mars is 3.3° south of Pluto. (Mars: 85% illuminated, magnitude 0.2, diameter 8.0". Pluto: magnitude 14.2. Sagittarius, evening sky.)

October 19th

08:55 UT – The waning gibbous Moon is north of Aldebaran. (Taurus, pre-dawn sky.)

October 21st

11:29 UT – Mercury is 3.6° north of Spica. (50% illuminated, magnitude -1.3, diameter 4.8". Virgo, not visible.)

The Orionid meteor shower peaks. Maximum zenith hourly rate: 25. (Moon: almost last quarter. Orion.)

October 22nd

08:45 UT – Dwarf planet Ceres reaches its maximum brightness. (Magnitude 7.2. Cetus, visible all night.)

13:29 UT – The almost last quarter Moon is south of Pollux (Gemini, pre-dawn sky.)

19:13 UT - Last Quarter Moon (Cancer, pre-dawn sky.)

22:54 UT – Dwarf planet Ceres is at perigee. (Magnitude 7.2. Cetus, visible all night.)

Asteroid 4 Vesta brightens to magnitude 7.5. (Cancer, pre-dawn sky.)

October 23rd

13:29 UT – The just-past last quarter Moon is south of M44, the Praesepe open star cluster. (Cancer, pre-dawn sky.)

October 24th

Venus leaves Scorpius and enters Ophiuchus. (80% illuminated, magnitude -4.0, diameter 13.4", evening sky.)

October 25th

00:29 UT – Dwarf planet Ceres is at opposition. (Magnitude 7.2. Cetus, visible all night.)

03:10 UT – The waning crescent Moon is south of Regulus. (Leo, pre-dawn sky.)

Neptune fades to magnitude 7.9. (Diameter 2.3". Aquarius, evening sky.)

October 26th

00:00 UT – Venus is 3.2° north of Antares. (Venus: 79% illuminated, magnitude -4.0, diameter 13.6". Ophiuchus, evening sky.)

October 27th

10:39 UT – Mercury reaches its maximum brightness. (100% illuminated, magnitude -1.4, diameter 4.7". Virgo, not visible.)

16:59 UT – Mercury is at superior conjunction with the Sun. Distance to Earth: 1.430 AU. (100% illuminated, magnitude -1.4, diameter 4.7". Virgo, not visible.)

Good opportunity to see Earthshine on the waning crescent Moon. (Pre-dawn sky.)

October 28th

10:01 UT – The waning crescent Moon is north of Jupiter. (Jupiter: magnitude -1.7, diameter 31.1", Virgo, pre-dawn sky.)

October 29th

08:59 UT – The almost new Moon is north of Spica. (Virgo, not visible.)

14:12 UT – Mars is at perihelion. Distance to Sun: 1.381 AU. (Mars: 86% illuminated, magnitude 0.3, diameter 7.6". Sagittarius, evening sky.)

Mercury leaves Virgo and enters Libra. (100% illuminated, magnitude -1.3, diameter 4.7", not visible.)

Asteroid 2 Pallas fades to magnitude 9.5. (Aquarius, evening sky.)

October 30th

08:19 UT – Venus is 3.0° south of Saturn. (Venus: 78% illuminated, magnitude -4.0, diameter 13.9". Saturn: magnitude 0.6, diameter 15.3". Ophiuchus, evening sky.)

17:38 UT - The second of two New Moons for October. The first occurred on the 1st. (Libra, not visible.)

20:40 UT – The new Moon is north of Mercury. (Mercury: 100% illuminated, magnitude -1.2, diameter 4.6". Libra, not visible.)

The Sun leaves Virgo and enters Libra.

October 31st

05:23 UT – Venus is at aphelion. Distance to Sun: 0.728 AU (Venus: 78% illuminated, magnitude -4.0, diameter 14.0". Ophiuchus, evening sky.)

November

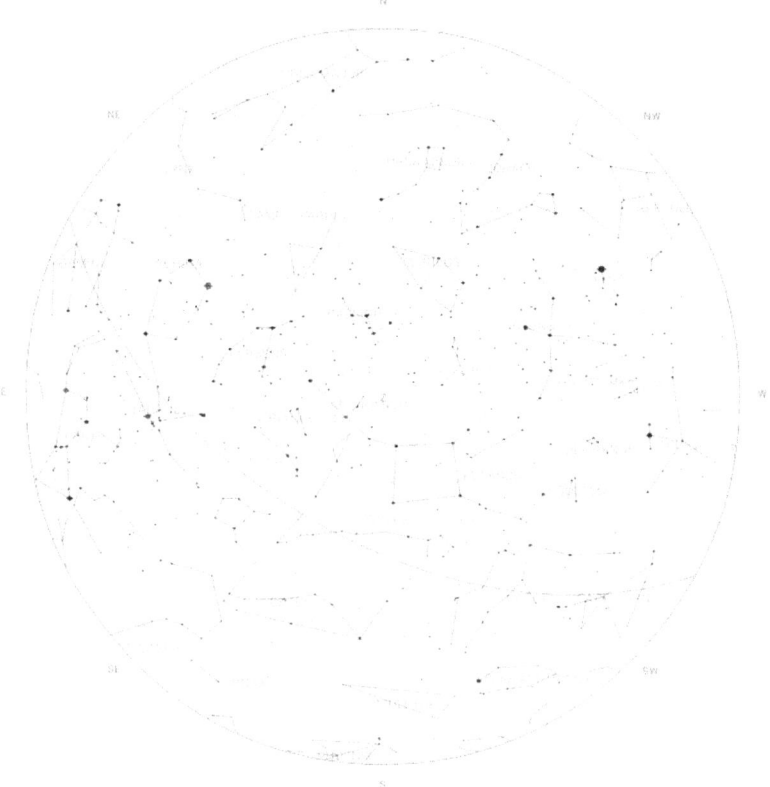

This image depicts the sky at 11:00 p.m. (Daylight Savings Time) on the 1st but the positions of the stars will be the same at 9:00 p.m. on the 15th and 8:00 p.m. on the 30th.

Phases of the Moon

First Quarter	Full Moon	Last Quarter	New Moon
November 7th	November 14th	November 21st	November 29th

November 2nd

05:48 UT – The waxing crescent Moon is north of Antares. (Ophiuchus, evening sky.)

18:00 UT – Mercury is at apogee. Distance to Earth: 1.441 AU. (Mercury: 100% illuminated, magnitude -1.0, diameter 4.6". Libra, not visible.)

21:06 UT – The waxing crescent Moon is north of Saturn. (Saturn: magnitude 0.5, diameter 15.3", Ophiuchus, evening sky.)

Mercury fades to magnitude -1.0. (100% illuminated, diameter 4.6". Libra, not visible.)

Saturn fades to magnitude 0.5. (Apparent diameter 15.3", Ophiuchus, evening sky.)

November 3rd

03:38 UT – The waxing crescent Moon is north of Venus. (Venus: 77% illuminated, magnitude -4.0, diameter 14.2". Ophiuchus, evening sky.)

Good opportunity to see Earthshine on the waxing crescent Moon. (Evening sky.)

November 5th

08:05 UT – The waxing crescent Moon is north of Pluto. (Pluto: magnitude 14.2. Sagittarius, evening sky.)

November 6th

10:16 UT – The almost first quarter Moon is north of Mars. (Mars: 86% illuminated, magnitude 0.4, diameter 7.3". Sagittarius, evening sky.)

November 7th

19:51 UT - First Quarter Moon (Aquarius, evening sky.)

November 8th

Mars leaves Sagittarius and enters Capricornus. (86% illuminated, magnitude 0.4, diameter 7.2", evening sky.)

Asteroid 3 Juno leaves Libra and enters Scorpius. (Magnitude 11.3, not visible.)

November 9th

13:08 UT – The waxing gibbous Moon is north of Neptune. (Neptune: magnitude 7.9, diameter 2.3". Aquarius, evening sky.)

Venus leaves Ophiuchus and enters Sagittarius. (76% illuminated, magnitude -4.0, diameter 14.6", evening sky.)

Mars fades to magnitude 0.5. (86% illuminated, diameter 7.2". Capricornus, evening sky.)

November 11th

15:02 UT – Mercury is at aphelion. Distance to Sun: 0.4677 AU. (97% illuminated, magnitude -0.7, diameter 4.7". Libra, not visible.)

November 12th

10:41 UT – The waxing gibbous Moon is south of Uranus. (Uranus: magnitude 5.7, diameter 3.7". Pisces, evening sky.)

Mercury leaves Libra and enters Scorpius. (97% illuminated, magnitude -0.7, diameter 4.7", not visible.)

Venus increases its apparent diameter to 15.0". (75% illuminated, magnitude -4.1. Sagittarius, evening sky.)

November 14th

13:52 UT - Full Moon. This is the largest full moon of the year, the largest for the past 68 years and the largest for the next 18 years. Apparent diameter: 33.506'. (Taurus, visible all night.)

20:49 UT – The full Moon is south of M45, the Pleiades open star cluster. (Taurus, visible all night.)

November 15th

15:45 UT – The just-past full Moon is north of Aldebaran. (Taurus, visible all night.)

November 17th

Mercury leaves Scorpius and enters Ophiuchus. (95% illuminated, magnitude -0.5, diameter 4.9", not visible.)

Mercury fades to magnitude -0.5. (95% illuminated, diameter 4.9". Ophiuchus, not visible.)

The Leonid meteor shower peaks. Maximum zenith hourly rate: 15. (Moon: waning gibbous. Leo.)

November 18th

17:11 UT – Mercury is 2.8° north of Antares. (Mercury: 94% illuminated, magnitude -0.5, diameter 4.9". Ophiuchus, not visible.)

17:51 UT – The waning gibbous Moon is south of Pollux. (Gemini, pre-dawn sky.)

November 19th

17:28 UT – The waning gibbous Moon is south of M44, the Praesepe open star cluster. (Cancer, pre-dawn sky.)

Dwarf planet Ceres fades to magnitude 7.5. (Cetus, evening sky.)

November 20th

06:40 UT – Neptune is stationary prior to resuming prograde motion. (Neptune: magnitude 7.9, diameter 2.3". Aquarius, evening sky.)

Mercury increases its apparent diameter to 5.0". (93% illuminated, magnitude -0.5. Ophiuchus, not visible.)

November 21st

08:33 UT - Last Quarter Moon (Leo, pre-dawn sky.)

11:57 UT – The last quarter Moon is south of Regulus. (Leo, pre-dawn sky.)

November 22nd

The Sun leaves Libra and enters Scorpius.

November 23rd

Saturn decreases its apparent diameter to 15.0". (Magnitude 0.5, Ophiuchus, evening sky.)

Asteroid 3 Juno leaves Scorpius and enters Ophiuchus. (Magnitude 11.2, not visible.)

November 24th

00:33 UT - Mercury is 3.4° south of Saturn. (Mercury: 91% illuminated, magnitude -0.5, diameter 5.1". Saturn: magnitude 0.5, diameter 15.0". Ophiuchus, not visible.)

22:44 UT – Venus is 3.5° south of Pluto. (Venus: 71% illuminated, magnitude -4.1, diameter 16.1". Pluto: magnitude 14.3. Sagittarius, evening sky.)

Asteroid 4 Vesta brightens to magnitude 7.0. (Cancer, pre-dawn sky.)

November 25th

00:16 UT – The waning crescent Moon is north of Jupiter. (Jupiter: magnitude -1.8, diameter 32.4", Virgo, pre-dawn sky.)

16:52 UT – The waning crescent Moon is north of Spica. (Virgo, pre-dawn sky.)

November 26th

04:33 UT – Asteroid 3 Juno is at apogee. Distance to Earth: 4.303 AU. (Magnitude 11.2. Ophiuchus, not visible.)

Good opportunity to see Earthshine on the waning crescent Moon. (Pre-dawn sky.)

November 28th

07:46 UT - Summer begins in the southern hemisphere of Mars. (88% illuminated, magnitude 0.6, diameter 6.6". Capricornus, evening sky.)

November 29th

12:18 UT - New Moon (Ophiuchus, not visible.)

14:40 UT – The new Moon is north of Antares. (Ophiuchus, not visible.)

The Sun leaves Scorpius and enters Ophiuchus.

November 30th

05:48 UT – The just-past new Moon is north of Saturn. (Saturn: magnitude 0.5, diameter 15.0", Ophiuchus, not visible.)

Mercury leaves Ophiuchus and enters Sagittarius. (85% illuminated, magnitude -0.5, diameter 5.4", evening sky.)

December

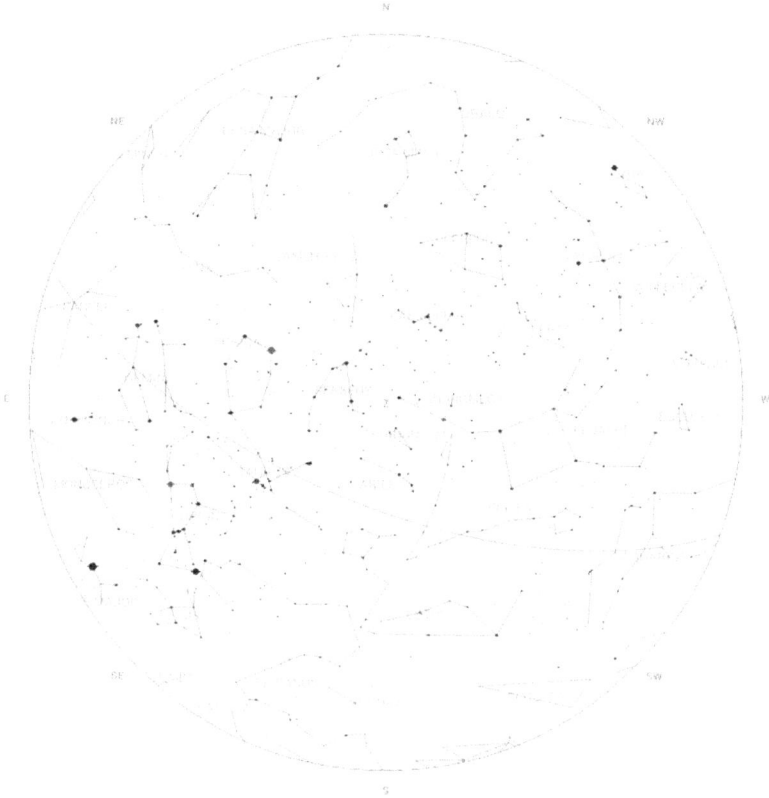

This image depicts the sky at 10:00 p.m. on the 1st but the positions of the stars will be the same at 9:00 p.m. on the 15th and 8:00 p.m. on the 31st.

Phases of the Moon

First Quarter	Full Moon	Last Quarter	New Moon
December 7th	December 14th	December 21st	December 29th

December 1st

02:56 UT – The waxing crescent Moon is north of Mercury. (Mercury: 83% illuminated, magnitude -0.5, diameter 5.5". Sagittarius, evening sky.)

19:34 UT – Asteroid 3 Juno is in conjunction with the Sun. Distance from Sun: 3.333 AU. (Juno: magnitude 11.2, Ophiuchus, not visible.)

December 2nd

03:13 UT – Asteroid 4 Vesta is stationary prior to beginning retrograde motion. (Magnitude 6.9. Cancer, pre-dawn sky.)

18:13 UT – The waxing crescent Moon is north of Pluto. (Pluto: magnitude 14.3. Sagittarius, evening sky.)

21:36 UT – Asteroid 3 Juno reaches its minimum brightness. (Magnitude 11.2. Ophiuchus, not visible.)

Good opportunity to see Earthshine on the waxing crescent Moon. (Evening sky.)

December 3rd

10:50 UT – The waxing crescent Moon is north of Venus. (Venus: 68% illuminated, magnitude -4.2 diameter 17.1". Sagittarius, evening sky.)

December 5th

08:56 UT – The waxing crescent Moon is north of Mars. (Mars: 88% illuminated, magnitude 0.7, diameter 6.4". Capricornus, evening sky.)

12:37 UT – Mercury reaches maximum brightness. (76% illuminated, magnitude -0.5, diameter 5.9". Sagittarius, evening sky.)

December 6th

23:52 UT – The almost first quarter Moon is north of Neptune. (Neptune: magnitude 7.9, diameter 2.2". Aquarius, evening sky.)

Venus leaves Sagittarius and enters Capricornus. (67% illuminated, magnitude -4.2, diameter 17.4", evening sky.)

December 7th

09:03 UT - First Quarter Moon (Aquarius, evening sky.)

December 9th

19:08 UT – The waxing gibbous Moon is south of Uranus. (Uranus: magnitude 5.7, diameter 3.6". Pisces, evening sky.)

December 10th

07:48 UT – Saturn is at apogee. Distance to Earth: 11.031 AU. (Saturn: magnitude 0.5, diameter 15.0", Ophiuchus, not visible.)

13:23 UT – Saturn reaches its minimum brightness. (Saturn: magnitude 0.5, diameter 15.0", Ophiuchus, not visible.)

14:55 UT – Saturn is in conjunction with the Sun. Distance to Earth: 11.031 AU. (Saturn: magnitude 0.5, diameter 15.0", Ophiuchus, not visible.)

December 11th

04:28 UT – Mercury reaches Greatest Eastern Elongation. (62% illuminated, magnitude -0.4, diameter 6.6". Sagittarius, evening sky.)

Uranus fades to magnitude 5.8. (Apparent diameter 3.6". Pisces, evening sky.)

December 12th

09:37 UT – The waxing gibbous Moon is south of M45, the Pleiades open star cluster. (Taurus, evening sky.)

December 13th

05:50 UT – The almost full Moon is north of Aldebaran. (Taurus, visible all night.)

The Geminid meteor shower peaks. Maximum zenith hourly rate: 120. (Moon: almost full. Gemini.)

December 14th

00:05 UT - Full Moon. This is the northernmost full moon of the year. (Taurus, visible all night.)

03:16 UT – Dwarf planet Ceres is stationary prior to resuming prograde motion. (Magnitude 7.8. Cetus, evening sky.)

16:48 UT – Mercury is 50% illuminated. (Magnitude -0.3, diameter 7.2". Sagittarius, evening sky.)

December 15th

Mars leaves Capricornus and enters Aquarius. (89% illuminated, magnitude 0.8, diameter 6.1", evening sky.)

December 16th

06:24 UT – The waning gibbous Moon is south of Pollux. (Gemini, pre-dawn sky.)

December 17th

03:32 UT – The waning gibbous Moon is south of M44, the Praesepe open star cluster. (Cancer, pre-dawn sky.)

05:00 UT - Mercury fades to magnitude 0.0. (40% illuminated, diameter 7.8". Sagittarius, evening sky.)

The Sun leaves Ophiuchus and enters Sagittarius.

December 18th

17:37 UT – The waning gibbous Moon is south of Regulus. (Leo, pre-dawn sky.)

December 19th

06:44 UT - Mercury is stationary prior to beginning retrograde motion. (30% illuminated, magnitude 0.4, diameter 8.2". Sagittarius, evening sky.)

Mercury fades to magnitude 0.5. (31% illuminated, diameter 8.2". Sagittarius, evening sky.)

December 21st

01:55 UT - Last Quarter Moon (Virgo, pre-dawn sky.)

10:44 UT - Winter Solstice. Winter begins in the northern hemisphere, summer begins in the southern hemisphere.

Mercury fades to magnitude 1.0. (21% illuminated, diameter 8.7". Sagittarius, evening sky.)

December 22nd

17:30 UT – The just-past last quarter Moon is north of Jupiter. (Jupiter: magnitude -1.9, diameter 34.6", Virgo, pre-dawn sky.)

20:48 UT – The just-past last quarter Moon is north of Spica. (Virgo, pre-dawn sky.)

Venus increases its apparent diameter to 20.0". (77% illuminated, magnitude -4.3. Capricornus, evening sky.)

December 23rd

Asteroid 4 Vesta brightens to magnitude 6.5. (Cancer, pre-dawn sky.)

The Ursid meteor shower peaks. Maximum zenith hourly rate: 10. (Moon: waning crescent. Ursa Minor.)

December 25th

14:40 UT – Mercury reaches perihelion. Distance to Sun: 0.308 AU. (5% illuminated, magnitude 3.1, diameter 9.6". Sagittarius, not visible.)

Good opportunity to see Earthshine on the waning crescent Moon. (Pre-dawn sky.)

December 26th

20:39 UT – The waning crescent Moon is north of Antares. (Ophiuchus, pre-dawn sky.)

Jupiter increases its apparent diameter to 35.0". (Magnitude -1.9, Virgo, pre-dawn sky.)

December 27th

21:24 UT – The waning crescent Moon is north of Saturn. (Saturn: magnitude 0.5, diameter 15.1", Ophiuchus, pre-dawn sky)

Dwarf planet Ceres fades to magnitude 8.0. (Cetus, evening sky.)

December 28th

18:41 UT – Mercury is at inferior conjunction with the Sun. (1% illuminated, magnitude 4.9, diameter 9.9". Sagittarius, not visible.)

December 29th

03:37 UT – The almost new Moon is north of Mercury. (1% illuminated, magnitude 4.8, diameter 9.9". Sagittarius, not visible.)

06:53 UT - New Moon (Sagittarius, not visible.)

07:13 UT – Mercury is at aphelion. Distance to Earth: 0.674 AU. (1% illuminated, magnitude 4.7, diameter 9.9". Sagittarius, not visible.)

14:26 UT – Uranus is stationary prior to resuming prograde motion. (Uranus: magnitude 5.8, diameter 3.6". Pisces, evening sky.)

December 30th

00:29 UT – The just-past new Moon is north of Pluto. (Pluto: magnitude 14.3. Sagittarius, not visible.)

Uranus decreases its apparent diameter to 3.5". (Magnitude 5.8, Pisces, evening sky.)

December 31st

Venus leaves Capricornus and enters Aquarius. (57% illuminated, magnitude -4.3, diameter 21.7", evening sky.)

www.ingramcontent.com/pod-product-compliance
Lightning Source LLC
Chambersburg PA
CBHW051548170526
45165CB00002B/924